KU-341-622

Environmental Dilemmas

Environmental Dilemmas

Ethics and decisions

Edited by

R.J. Berry FRSE
Department of Biology
University College London
London, UK

CHAPMAN & HALL
London · Glasgow · New York · Tokyo · Melbourne · Madras

Published by Chapman & Hall, 2–6 Boundary Row, London SE1 8HN

Chapman & Hall, 2–6 Boundary Row, London SE1 8HN, UK

Blackie Academic & Professional, Wester Cleddens Road, Bishopbriggs, Glasgow G64 2NZ, UK

Chapman & Hall, 29 West 35th Street, New York NY10001, USA

Chapman & Hall Japan, Thomson Publishing Japan, Hirakawacho Nemoto Building, 6F, 1–7–11 Hirakawa-cho, Chiyoda-ku, Tokyo 102, Japan

Chapman & Hall Australia, Thomas Nelson Australia, 102 Dodds Street, South Melbourne, Victoria 3205, Australia

Chapman & Hall India, R. Seshadri, 32 Second Main Road, CIT East, Madras 600 035, India

First edition 1993

Typeset in 10/12 pt Palatino by Keyboard Services, Luton, Bedfordshire

Printed in Great Britain by T.J. Press (Padstow) Ltd, Padstow, Cornwall

ISBN 0 412 39800 1

A catalogue record for this book is available from the British Library

Library of Congress Cataloging-in-Publication data available

∞

Printed on permanent acid-free text paper, manufactured in accordance with the proposed ANSI/NISO Z 39.48–199X and ANSI Z 39.48–1984

Contents

Contributors

Lord Ashby, FRS,
Clare College, Cambridge CB2 1TL.

R.J. Berry, FRSE,
Department of Biology, University College London, Gower Street, London WC1E 6BT.

A.A. Brennen,
Department of Philosophy, University of Western Australia, Perth WA 6009, Australia.

P. Brimblecombe,
Scholl of Environmental Sciences, University of East Anglia, Norwich NR4 7TJ.

D.A. Everest,
UK Centre for Economic and Environmental Development, 3 King's Parade, Cambridge CB2 1SJ.

F.B. Golley,
Institute of Ecology, University of Georgia, Athens GA 30602, USA.

B.H. Green,
Department of Agriculture, Horticulture and the Environment, Wye College, Kent TN25 5AH.

W. Haber,
Lehrstuhl für Landschaftsokologie, Technische Universitat Munchen, D–8050 Freising 12, Germany.

R. Harrison,
34 Holland Park Road, London W14.

O.W. Heal,
Institute of Terrestrial Ecology, Bush Estate, Peniciuk, Midlothian EH 26 0QB.

J. Morton Boyd, CBE, FRSE,
57 Hailes Gardens, Edinburgh EH13 0JH.

B. Moss,
Department of Environmental and Evolutionary Biology, University of Liverpool, Liverpool L69 3BX.

F.N. Nicholas,
School of Environmental Sciences,
University of East Anglia, Norwich
NR4 7TJ.

J.C. Powell,
Environmental Appraisal Group,
School of Environmental Sciences,
University of East Anglia, Norwich
NR4 7TJ

L.E.J. Roberts, CBE, FRS,
School of Environmental Sciences,
University of East Anglia, Norwich
NR4 7TJ.

T.M. Roberts,
Institute of Terrestrial
Ecology, Monks Wood
Experimental Station,
Huntingdon, Cambridgeshire
PE17 2LS.

J. Sheail,
Institute of Terrestrial Ecology,
Monks Wood Experimental
Station, Huntingdon,
Cambridgeshire PE17 2LS.

R.K. Turner,
Environmental Appraisal Group,
School of Environmental Sciences,
University of East Anglia, Norwich
NR4 7TJ.

G. Wyburd,
International Chamber of
Commerce, ICC United Kingdom,
14–15 Belgrave Square, London
SW1X 8PS.

Preface

This is a book by people who have had to make decisions which affect the environment in which we all live, decisions which sometimes affect the quality of life of millions. It is not an academic disquisition on how to approach decision-making. Most of the chapters are written by scientists who have had to take action or make recommendations on environmental matters in situations where the data are incomplete or choices hedged by factors beyond scientific resolution; the result is that they have had to resolve dilemmas about the proper way forward in the matter. My brief to the authors was to describe issues with which they had been personally concerned, rather than simply select from the vast range of environmental problems 'out there'. The only exception to this was Andrew Brennan (Chapter 1), who is a professional philosopher; I asked him to say something about the processes and errors indulged by environmental decision-makers.

There is some overlap between chapters, but this is not extensive. I have made no attempt to eliminate it, because the aim has been to present personal points of view, not a systematic account of environmental problems. Similarly, there are important topics which are not covered. Indeed, a critic would complain that a book on environmental dilemmas which does not deal directly with the crucial divide between development and conservation is almost wholly irrelevant; from one point of view, it could be condemned as fiddling while Rome burns.

I make no apology for the selection of dilemmas represented herein. First, I believe that we are bound morally to espouse both development and conservation (which is not the same as vaguely aiming for growth and preservation at the same time); and second, virtually all major choices affecting the environment depend on or produce the sort of problems faced by the contributors of this book. In other words, the range of dilemmas opened up in the pages that follow are typical of those that recur in the real world.

One criticism that will be levelled at this book is that it is unbalanced; that nuclear dilemmas are described by a believer in nuclear power, animal welfare by a 'moderate', acid precipitation by establishment scientists, industry's dilemmas by an industrialist, and so on. I do not believe this is necessarily a weakness. All the writers set out the factors involved in their involvement, and show how they worked through their frustrations and contradictions. It would, of course, have been possible to produce the sort of book where different viewpoints are presented on the same subject. But there are plenty of in-depth studies of particular topics. I prefer to interpret the case-studies described herein as showing that environmental problems are not inevitably confrontational; that consensus on opinions as well as facts is not the same as compromise; that commitment should complement truth and not conflict with it.

I hope this collection of studies will help those facing environmental decisions, whether personal or institutional ones. I am grateful to all those who have contributed their experiences to this book. Their efforts will have been worthwhile if they help environmental dilemmas to be faced more immediately and efficiently than is too often the case.

R.J. Berry
September 1992

Foreword

Eric Ashby

Lord Ashby was Professor of Botany, Sydney University 1938–46 and Manchester University 1946–50; Vice-chancellor, Queen's University, Belfast 1950–59 and University of Cambridge 1967–69. He was the first Chairman of the Royal Commission on Environmental Pollution 1970–73; from 1973–84 he was closely involved in debates and committees on environmental policy in the House of Lords. He is the author of Reconciling Man with the Environment *(Oxford University Press, 1978),* The Search for an Environmental Ethic *(in* The Tanner Lectures on Human Values, *Cambridge University Press, 1980), and the* Politics of Clean Air *(with Mary Anderson) (Clarendon Press, 1981).*

One expects a foreword to be put first and written last. I write this foreword without a detailed knowledge of the contents of the book. No matter, for one word in the title gives away the book's theme: *dilemma*. A dilemma is defined as a 'position that leaves only a choice between two equally unwelcome possibilities'. The word aptly describes many of the choices governments have to make in the 1990s. When the city of Winnipeg celebrated its centennial in 1974 it held a great international symposium. The theme was not (as one might have expected) self-congratulation on past achievements. Speakers were asked to discuss 'Dilemmas of Modern Man'; how to survive what Alvin Toffler, a Cassandra of the 1960s, described as a 'fantastic compression of history' (Beamish, 1975).

The twentieth century is ending in an epidemic of environmental dilemmas. This book presents case histories of some of them; so, as a prologue, it may be useful to consider how governments set about making choices in environmental policy.

Individuals have to sort out some environmental dilemmas for themselves: to forgo the solace of a cigarette or to risk cancer; to expose oneself to a risk of *Salmonella* or to put up with the dry texture of a hard boiled egg; to vote at local elections for an increase in charges or to accept a decline in care for the city amenities. The guide to making such personal choices as these is assumed to be self-interest. But most environmental dilemmas cannot be solved so simply. It cannot be assumed that the preferences of individuals can be combined to produce a preference-statement for the whole society; indeed, the 'tragedy of the commons' is that unrestrained freedom for someone to maximize his own welfare (as economists put it) may put at risk the welfare of the whole group. This dilemma was vividly described by Garrett Hardin (1968). When animals are grazed on a common pasture each herdsman is tempted to increase the size of his flock. If this desire is not restrained the common will become overgrazed and ultimately destroyed; and all herdsmen using the common will suffer. But self-restraint on the part of any one herdsman will not save the common. It will merely impoverish the altruistic herdsman. The common can be preserved only by 'mutual coercion, mutually agreed upon by the majority of people affected'. This is but one example of a profound principle in ecology. There are many other such tragedies; all 'common property resources' are at risk. In the commons of the oceans some fisheries are already faced with disaster. In the commons of the air over Los Angeles 'the smog is the result of ten million individual pursuits of private gratification. But there is absolutely nothing any individual can do to stop its spread'. How to achieve 'mutual coercion, mutually agreed' is a problem that was tackled by philosophers long before it interested ecologists; by Thomas Hobbes, for instance, in the 17th century and Rousseau in the 18th century. It has not yet been solved.

So, dilemma-solving for the State is a different ball game from personal dilemma-solving. Over the last 150 years industrial countries have come to realize that the State must intervene to protect the environment from hazards due to human beings. Intervention almost always poses dilemmas. A decision to preserve a city's green belt might be welcomed by conservationists but would be unlikely to please developers; laws to control animal experiments would not satisfy the animal-rights lobby; bans on pesticides would anger farmers. Should the politician, out of superior knowledge and education, prescribe which of two incompatible 'goods' should be permitted by the State? If so, he runs into the hazard of collectivism, so emphatically criticized by A.V. Dicey (1914): the notion 'that the wish of the people may be overruled for the good of the people'. Or should he act as a sort of seismograph, recording public opinion, waiting for a strong enough reading on the political Richter Scale before he takes action? What advice does the politician seek and how does he use the advice?

Environmental dilemmas cannot be understood, let alone solved, without technical expertise. The ingredients for making an environmental choice come from half a dozen different disciplines. It is the task of the politician to blend the ingredients into a dish which parliament will digest. He has no expertise in the disciplines he draws upon. The qualification he can aspire to – it is a rare one – is a flair for integrating expert advice into acceptable legislation. History will judge the quality of his decisions by the tacit ethical principles – even though he rated them no higher than hunch – that guided his choices.

The first expert to be called in is likely to be the scientist. The trouble about his advice is that it is liable to be distorted once it gets into the political machinery, not through wilful falsification but through neglect of the small print. Politicians, adept as some of them are at making evasive pronouncements, dislike receiving evasive evidence; and it is inevitable that scientific evidence on complex issues, such as global warming, should be hedged with reservations and blurred by words like 'probably' and 'possibly'. 'Certainty' is not a word scientists like to use. They wince when they hear a Minister, after taking the best scientific advice, announce that some food can be regarded as *absolutely* safe to eat. Senator Muskie, in the USA, spoke for many politicians when he called for 'one armed' scientists; advisers who will not say 'On the one hand the evidence is so, but on the other hand . . .' (David, 1975).

One must not forget another difficulty in getting scientists and politicians onto the same wave-length. The whole ethos of science rests on the belief that there can be no compromise over scientific concepts. Two opposing hypotheses remain opposed until one or the other is discarded. Meanwhile judgement is suspended, however long it takes. Politics cannot be pursued that way. Compromise is generally the only way to reconcile opposing opinions and there is always a dateline by which a decision has to be made. This difference in style of thinking weakens the influence that scientists can have on environmental policy.

There is a third difficulty about acceptance of scientific advice by politicians. Some of the advice rests upon statistical calculations of risk. These may be accurate and based on trustworthy evidence (as are, for example, the risks of death in road and railway accidents.) But, as everyone knows, apprehension about a risk may have little to do with the statistical chance that the risky event will occur. For some risks no resort to reason seems able to reconcile the subjective perception with the calculated probability. Thus most of the 5182 deaths on Britain's roads in 1988 were reported as single incidents, scattered in time and location. A few lines in the local press was as much publicity as most of them received. But an air crash, killing one hundredth of that number, would be on television screens and newspaper headlines across the world. Statistical risk is compounded of frequency and severity. Perceived risk gives a higher weighting to severity than to frequency; it is influenced, too, by propinquity, and of course by the perceiver's responsibility for taking the risk: climbers and motor cyclists put a discount on the risks they take which they would not put upon the risk of a coach accident.

The politician, therefore, cannot act solely upon the statistical estimates of risks, even if he is convinced by them. It is the subjective, perceived, risk that he must be seen to be allaying, for this politically is more 'real'. Hence the pussyfooting in parliament over the compulsory use of seat-belts in cars (a compulsion regarded by some people as an infringement of civil liberty) despite evidence that the use of belts would save some 128 serious injuries or deaths on the roads of Britain every week (Ashby, 1979). In the whole world it is estimated that about a quarter of a million people die and ten million are injured as a result of road accidents each year (Sabey and Taylor, 1980). We are as reconciled to this as our great-grandparents were reconciled to deaths from typhoid.

There are irrationalities in the opposite direction too, when people over-react to negligible risks. A few years ago some Canadian pathologists produced tumours in rats by feeding them with saccharin. The dose was 2500 milligrams per day. The analogous dose for humans would be to drink 800 twelve-ounce bottles of soft drink per day. But their observation provoked a vigorous demand to ban saccharin and so, possibly – to hazard a guess – putting at risk some of the 50 million people in North America who were relying on saccharin to protect themselves from getting too fat.

So it should be no surprise that scientists who give advice about environmental dilemmas sometimes find that their advice sinks without trace.

Another expert called in to advise on environmental dilemmas is the economist. Economists seem less prone than scientists to qualify their advice and it is sometimes suspected that politicians are being told not only how to reach a goal but what goal to reach. To give economists their due, they do not (if they are sticking to their last) moralize about the way preferences ought to be made; what they do is to point out the likely logical outcome of holding certain preferences. Thus, pollution is no more 'bad' to an economist than cyanide is 'bad' to a chemist; both can tell the politician what harm can be done by it and what controls can be applied.

The economist's important contribution to environmental policy has been to point out, and to attempt to measure, the cost to society for the hitherto uncosted use of air, water, and land, exploited for human activities. Just as mass and velocity are convenient units for measurement in mechanics, so money is the economist's convenient unit for measurement. It is in fact, a kind of reductionism and it is prone to all the limitations and deceptions of reductionism, albeit more dangerous when it gets into political arguments; for mass and velocity are arcane words not familiar to the public, while everybody knows what money is. Simplifications that economists take for granted when they talk to one another get lost when their advice is incorporated into reports and memoranda. Three of these simplifications enter deeply into most environmental dilemmas and all three are unreliable as a guide for political choice. They are: (i) in making choices, people behave rationally; (ii) rational choice is based on self-interest; (iii) even objects not in the market place for sale (for example, Westminster Abbey or your wife/husband) have a money value. (For your wife/husband, in any case, the value has already been calculated and used to resolve certain environmental dilemmas, such as the cost of improving a dangerous stretch of road. Against this cost can be set the cost of accidents if the road isn't improved; this includes an estimate of the value of a human life and the number of lives that might be saved. The most recent 'notional cost of a road death' was about half a million pounds (Southwood, 1990), and if this seems generous I had better add that the trade-off for Americans and Canadians is even higher.)

For a discipline devoted to the logic of choice, simplifications of this sort are convenient, but for the choice that has to be made in many environmental dilemmas they offer no safe guidance. Under section 9 of the Wildlife and Countryside Act, Parliament protects the natterjack toad, which very few MPs or their constituents have ever seen, or would recognize if they did see. The even more modest Furbish Lousewort held up the building of a dam in Maine, on the orders of the president of the United States under the US Endangered Species Act (Ashby, 1980). It is hard to identify these as rational choices based on collective self-interest. One has to fall back upon the belief that politicians make decisions like these in order to comply with unquantified and tacit values held by the citizens who voted them into office.

One should not be surprised if the jerk-reaction of a politician to an environmental dilemma is to tot up the votes that might be won or lost by the decision. I once had to wait upon a Minister to try to persuade him to tighten some regulations to control pollution into rivers. I went to his room well briefed with arguments to support my case. I had hardly started to unfold these arguments when he interrupted me. 'You are pushing at an open door' he said, 'Don't you know that more people fish on a Saturday afternoon than play football?'.

Anglers lobbying for clean rivers are acting out of self-interest. But this sentiment is not sufficient to account for the trend of increasing empathy for the welfare of what is sweepingly called the biosphere; nor does it account for the willingness of governments (sometimes in the face of the advice cost-benefit analysts would give) to protect endangered species.

At first sight it seems inconsistent with the doctrines of a market economy. Firms that can't make a profit are labelled lame ducks and left to suffer the fate of injured birds: they are wiped out by predators. It could be argued that endangered species are the lame ducks of evolution; a pity, but that's the way things are if habitats are destroyed. Have not the habitats of the horse-drawn van, the village blacksmith, the full-time cook and housemaid, been destroyed too?

But that is not the way we argue. At a time when politicians were facing plenty of dilemmas about the economy of the nation, it was moving to listen to the British parliament debating the Wildlife and Countryside Bill. In the House of Lords there were some 2000 amendments to the draft Bill. On one occasion there was a long discussion about bats. Clause 9 of the Bill read that it would be an offence if any person intentionally 'damages or destroys any structure or place' which any protected animal uses for 'shelter or protection'. But suppose bats squat in your bathroom? There is an escape clause, that nothing in clause 9 'shall make anything unlawful done within a dwelling house'. The peers were not satisfied with this. First, there was an amendment, to add to 'damages or destroys' the words 'or obstructs access to'. This would make it an offence to block the entrance to your barn or garage to keep out bats already in possession. The amendment passed, and stands in the Act (HMSO, 1981). But this was not sufficient for some people. Bats do like to occupy attics and basements of large houses which are also occupied by some peers; and when the debate was resumed the next day one peer moved that the lines exempting dwelling houses should be deleted. This was resisted; it was agreed that to offer hospitality to bats in your own living quarters was going too far, whereupon another peer proposed an amendment by way of compromise. It was to delete 'dwelling house' and to insert in its place:

> the occupied quarters of a dwelling house, which term shall not be taken to include the roof space, cellar, cavity-wall or vacuity in the fabric of a dwelling house that is not ordinarily used for dwelling purposes.

After more discussion, the amendment was dropped, but to evict bats from any other part of your property is an offence for which you might be fined up to £1000.

The Wildlife and Countryside Act did not please everybody, but – and this is the interesting point – not a single voice was raised to dismiss it as a fatuous waste of Parliament's time, totally at odds with the enterprise economy. It was evident from the long debates in both Houses that this sort of environmental policy is something the British people want.

From this and much other environmental legislation it is clear that the simple criteria of choice that economists use do not fully explain the choices made by governments on behalf of the people. These choices are surely 'what people want', or politicians would not take the trouble to make them as they do. So what is known about the ethical foundations of this tacit approval?

Not very much. Among intellectuals the word 'stewardship' is the best summary of ethics for the environment but I doubt whether popular approval articulates it that way. A century ago T.H. Huxley described the situation with his incisive clarity. Genetically, we are coded to protect self-interest, equipped to survive under a 'law of the jungle'. But in the community self-interest has to co-exist with society's interest. This puts constraints on behaviour. In time the constraints are codified as customs and ultimately as law, a law of the garden rather than a law of the jungle. Huxley put it this way (Huxley, 1925a):

> It becomes impossible to imagine some acts without disapprobation, or others without approbation of the actor, whether he be oneself, or anyone else. We come to think in the acquired dialect of morals.

It can be argued that to adopt the dialect of morals is only another form of self-interest. This may be true, but there is little difficulty in separating the two kinds. The self-interest that prompts one to buy a bottle of wine is palpably different from the self-interest that deters one from throwing the empty bottle into the street. As Huxley wrote: '. . . the greatest restrainer of the anti-social tendencies of men is fear, not of the law, but of the opinion of their fellows' (Huxley 1925b).

Stewardship? Fear of disapprobation from the neighbours? Or perhaps a surge of guilty conscience at the damage we have already done to nature? Whatever the cause, there is clear evidence of the effects. Consider the massive membership of voluntary organizations concerned with wildlife and the environment. It has been estimated that some three million people in Britain, one in nineteen of the population, belong to one or other of these organizations. What was, a few years ago, only an enthusiasm, has become a movement, courted by politicians. (It is a fitting symbol that the flag of the movement is green, suggesting chlorophyll, which links the sun's energy with life on earth.) It is the force of this movement, rather than advice from scientists and economists, that spurs political action. The success of the campaign to remove lead from petrol is an example of this. If politicians had followed scientific advice alone, removing lead from petrol would have been regarded as desirable, but not the top priority even for dealing with lead pollution. Pressure groups, amplified by hype in the media, gave the impression that lead in petrol was the chief source of lead pollution. In fact, among adults the percentage for the daily uptake of lead which can be attributed to petrol varies from 16% in a rural town to 39% in an inner city; serious and certainly needing action, but no more than the action needed for three or four million households whose water supply carries lead: in some areas with lead plumbing and plumbosolvent water, 50% cent of the total uptake of lead comes from the kitchen tap (RCEP, 1983). When the Royal Commission on Environmental Pollution published its ninth report (*Lead in the Environment*) in 1983 there was an astonishingly swift and positive response from the Government (p. 234). This swift response was certainly due to the lead-in-petrol campaign. To persist in using four-star petrol is already to invite what Huxley called 'disapprobation'. Indeed some motorists seek positive approbation by putting a sticker on the car window which reads:

I use unleaded. Can you?

To the dialect of morals a new and cumbersome compound word has been added: the mores of the 1990s require people to be *environment-friendly*. It is the test to apply to your car, central heating, refrigerator, fly-spray, even eggs and vegetables bought in the supermarket. All are manifestations of the environmental ethic.

Loyalty to this ethic is about to be tested in a dismaying challenge. Until recently, it could be said that the pursuit of self-interest in environmental dilemmas was kept in bounds (as Huxley wrote) by fear of disapprobation from the neighbours. This has been an influence also among the Governments in the European Community. But at the end of the 1980s a road-to-Damascus revelation fell upon industrial nations. They became convinced at last that activities highly beneficial to today's society are likely to bring disaster to some parts of the globe in the 21st century. The greenhouse effect and damage to the ozone layer do not threaten any person on the voting registers of present-day Europeans or Japanese or Americans. Yet the only way to diminish dangers ahead is to pay a high insurance premium now for the sole benefit of posterity. Moreover there is no guarantee that the insurance premium will suffice, nor do we know how far ahead the generation is that will suffer if we do not act now. It is a tribute to the political systems of several nations that encouraging statements of intent have been made, declarations that those responsible for damaging the biosphere will, in good time, refrain, and help developing nations to avoid making similar mistakes. If these good intentions are ever to be turned into deeds there will need to be an unprecedented resolve on the part of millions of people to forgo benefits and even to make sacrifices for the sake of a society they will never see; and there will need to be unprecedented courage on the part of popularly elected governments to satisfy electors who have not yet been born, at the risk of alienating electors whose votes put them into office.

Eric Ashby
Clare College
Cambridge

REFERENCES

Ashby, E. (1979) Reflections on the costs and benefits of environmental pollution, *Perspectives in Biology and Medicine*, **23**, 7–24.
Ashby, E. (1980) What price the Furbish Lousewort?, *Environmental Science and Technology*, **14**, 1176–81.
Beamish R.E. (ed.) (1975) *Dilemmas of Modern Man*. Great West Life Assurance Co., Winnipeg.
David, E.E. (1975) One armed scientists, *Science*, **189**, 891.
Dicey, A.V. (1914) *Law and Public Opinion in England*, Macmillan, London, p. lxxiii.
Hardin, G. (1968) The tragedy of the Commons, *Science*, **162**, 1243–8.
HMSO (1981) *Wildlife and Countryside Act 1981*, HMSO London.
Huxley, T.H. (1925a) *Evolution and Ethics*, Macmillan, London, p. 30.
Huxley, T.H. (1925b) *Evolution and Ethics*, Macmillan, London, p. 29.
Royal Commission on Environmental Pollution (RCEP) (1983) *Lead in the Environment*, ninth report, Cmnd 8852.
Sabey, B.E. and Taylor, H. (1980) The known risks we run, in *Societal Risk Assessment* (eds R.C. Schwing and W.A. Albers, Jr), Plenum Press, New York, p. 47.
Southwood, R. (1990) Risk in the natural world and human society, *Science and Public Affairs*, **5**, 85–99.

Environmental decision-making

1

Andrew A. Brennan

Andrew Brennan has moved recently to the Professorship of Philosophy at the University of Western Australia in Perth; until 1991 he was Reader in Philosophy at the University of Stirling, Scotland. He is the áuthor of Thinking about Nature: an Investigation of Nature, Value and Ecology *(Routledge and University of Georgia Press, 1988), a book which reveals a much greater awareness and understanding of scientific writings than most works by philosophers.*

1.1 INTRODUCTION

Decisions involve both the past and the future. For the individual who is facing a decision, the past operates as a precedent and the current decision adds in its turn to the stock of precedents to be called upon in future. These precedents are by no means binding. We all sometimes take decisions that are out of character; on other occasions, we wrestle with the problem of making a new start or breaking an old habit. Although we value consistency in our lives, we also try to innovate.

Taking decisions is a complex business and thinkers have wrestled with the problems of individual decision-taking for a very long time. In ancient times, Plato expressed the view that knowledge of the good was not

Environmental Dilemmas Ethics and decisions
Edited by R.J. Berry
Published in 1992 by Chapman & Hall, London. ISBN 0 412 39800 1

compatible with action contrary to the good. Wrongdoing and the associated bad decisions were thus a kind of ignorance. This led to an immediate problem for Aristotle*. If virtue consists of knowing what is good, desiring it and then acting consistently with it, there will be several ways of falling short of virtue. For example, we may know what is good, and act in accordance with this, although not truly desiring the good. This is a kind of controlled behaviour that, for Aristotle, stops short of the true virtue.

Worse possibilities arise. Someone may be ignorant of what is good, fail to desire the good and not act according to the good. Aristotle regarded this as a clear case of vice. But an intermediate case worried him and still puzzles contemporary moral philosophers. This is where we know what is good, desire that the good should occur, and yet act contrary to both what we know and desire. In the philosophical literature, this is usually described as the problem of *incontinence* (or in Greek, *akrasia*), although it is sometimes also referred to as *weakness of will*. Whatever label we use, the phenomenon is still very much with us.

1.2 WEAKNESS OF WILL AND ENVIRONMENTAL DEGRADATION

In many ways the consumer society makes it easy for us to be incontinent in Aristotle's sense. We know that energy-intensive manufacturing, over-packaging, introduction of new chemical agents and failure to recycle are serious failings in much of the industrialized world. Yet, despite this knowledge, and almost against our desires, we find ourselves sucked into consumption, choosing an item more for its attractive packaging or its image than for any real need it satisfies. Indeed, we often use the language of the weak-willed when we think of the ways our society allows us to *cosset* ourselves or *pander* to our whims. A society of agents who are cosseted and pandered to in this way is hardly the society of noble, rational, reflective, virtuous intellectuals held up as ideals by the Greeks.

Aristotle's hero was someone who set out to cultivate the virtues by seeking moderation in all things as a mean between the vices of excess. This is someone who is neither cowardly nor foolhardy, but brave; who neither denies bodily pleasure nor engages in self-indulgence, but is temperate. Aristotle was in no doubt that the social conditions of Athens made it possible for at least some of its citizens to aspire to such virtuous self-development. Unlike the Athenians of old, our society no longer has a slave caste and women have the same political entitlements as men.

* As an example see book VII of Aristotle's *Nicomachean Ethics* and the commentary in Ch. 17 of A. Edel (1982) *Aristotle and His Philosophy*, Chapel Hill, University of North Carolina Press.

Have we, in creating societies free of slavery and with a commitment to social equality, somehow lost the ideal of individual excellence?

This last question can be considered in a number of ways. We can take society itself as providing a number of constraints and possibilities for individual growth. It is then reasonable to wonder just what directions of development are favoured by particular kinds of social arrangement. For example, it may be that large consumer societies of anonymous purchasers discourage virtuous modes of behaviour. This line of argument is taken by some American writers who have compared modern consumer societies with earlier pioneering ones*. In the latter, they argue, with fewer pressures on the environment, and people organized into close-knit communities, the question 'How would you like it if somebody did that to you?' has real point. Where there are few agents, each decision's consequences are widely known, and agents are fully aware of the possibility that their roles may one day be exchanged. By contrast, the ethical *quid pro quo*, like the barter that characterized the early rural communities, seems no longer to be part of our everyday consciousness.

Such appeals to pioneer communities have an uncomfortable sense of the mythic. Early American communities were not particularly known for their sensitivity to waste handling (nor did the pioneers show particular admiration for the environmental sensitivity of the primary peoples they displaced). Taming and subduing the wilderness is not a project that would appeal to the modern conservationist in any case, and I am doubtful if we can learn many lessons about modern societies from such comparisons. What is true about such communities, however, is that cosseting and pandering were not particularly widespread features of life. Necessities had to be won by hard work and self-reliance was essential for survival. Hot baths no doubt assumed a special status when the fuel to heat the water had to be hewn from the forest by the axe.

In these circumstances, self-development involved different forms of activities from those afforded in contemporary industrialized culture. The modern pluralist society is not only host to a variety of points of view and community goals but also to a variety of ways of expressing individuality. This variety is perhaps one source of its environmental undoing. The very openness of the society puts it in a poor position to protect the basic features on which the agents within it depend for their continued existence. The pursuit of individual preference, in an intelligent but weak-willed way, has led to a situation in which all preference-seeking is threatened. Having come to such a pass, individual self-control (what Aristotle called 'continence') may not be enough to put matters right. Moreover, continence brings few rewards while weakness of will is nice

* See the editor's introduction to *Upstream/Downstream: Issues in Environmental Ethics* (1990) (ed. D. Scherer), Temple University Press, Philadelphia.

while it lasts. Suppose we agree that we ought to reform the way we live; suppose, further, that we agree that a mature, intelligent society would care about the situation of other, poorer societies and about the health of the biosphere on which all social life depends. As Aristotle recognized we can know what ought to be done, we can even desire to do it, and yet we find ourselves giving in to temptation. As he also recognized, it is not a phenomenon confined to stupid or uneducated people. There is little point, then, in bewailing the environmental stupidity of people in our society for that may be to mistake the nature of the problem.

1.3 CORPORATE ACTORS AND INDIVIDUAL RESPONSIBILITY

A purely philosophical treatment of decisions might stop at this point to consider the connections between incontinence, on the one hand, and self-deception on the other. Self-deception is a puzzling phenomenon, closely allied to incontinence. One tempting explanation for weakness of will is that the agent involved finds ways of deceiving him- or herself. We see that a course of action, for example, is against our interest; but somehow we talk ourselves into taking it anyway. If asked about it we deny fervently that the action is contrary to our interests. We seem both to know and not know something. Sartre referred to an allied phenomenon as 'bad faith', and – discerning it as widespread and largely unrecognized by its victims – he took it as evidence of the depressing non-authentic nature of our existence*. Proper self-development, for him, involved striving for greater honesty in both relationships with others and also our conception of ourselves and our situation.

It is hard for a philosopher familiar with the literature on incontinence and self-deception not to be reminded of these things when reading about environmental issues. Increasingly, groups like the International Chamber of Commerce, large corporations, and government bodies, produce analyses and claims which embody apparent self-deception. For example, it seems to be widely believed that poverty in the poor countries is a major cause of environmental degradation†. To alleviate this poverty it is allegedly necessary to engage in further industrial development which will, in currently vogue terms, be 'sustainable'. This new industrial growth will generate the resources to protect diminishing resources. The Brundtland aim of 'sustainable development' becomes, in these terms, no more than a commitment to business-as-usual with an increased emphasis on pollution control, energy-saving and other green matters.

In case the above remarks are thought to be an exaggeration, consider

* For a clear statement of Sartre's position, see the relevant chapter in Leslie Stevenson's (1990) *Seven Theories of Human Nature*, Oxford University Press, Oxford.
† This impression is given in the recent UK White Paper on the Environment – *This Common Inheritance*, HMSO, London, Cm 1200, Sections 4.4–4.12.

the following remarks by Peter Wallenberg, President of the International Chamber of Commerce:

> Broadly speaking, business and other proponents of a free enterprise economy believe that man's ingenuity will make sustainable growth possible and compatible with the interests of subsequent generations. Apart from other considerations, a growing world economy is essential to alleviate the often desperate poverty in many developing countries*.

Aside from the sexist language and the technological optimism of the above passage, it is interesting to note how the notion of sustainable development has been replaced with the idea of sustainable *growth*. The adoption of a common business and governmental approach to the problems of sustainable development takes little account of the past history of industrialization. The richest countries of the world are not renowned for their ability to maintain (say) the diversity of species present before industrialization. In the UK, for example, intensive agriculture and an expanding economy have led to significant reductions in virtually all our native fauna and flora. Moreover, although it is true that poverty in some countries is associated with environmental degradation, we need to consider the causes of that poverty. It is well known that there is a net outflow of resources every year from the poor to the rich countries: the global economic system is in many ways set up as if to exploit the resources of the poorer countries for the satisfaction of the rich ones.

It would be useful if these last points were discussed in the publications of large corporations and government bodies. This is, on the whole, not so. Despite a passing reference to debt for nature swaps, the UK government's 1990 White Paper on the environment does not go into any depth on the issue of global resource security and its relations to world poverty. In fact, as recent Worldwatch Institute figures show, the gap between the poorest and the richest nations has widened significantly over the past 40 years. Moreover, the very poorest nations have stayed, in absolute terms, almost as poor as they were 40 years ago (Fig. 1.1)†. The appeal to business as usual, like Wallenberg's confidence in 'a growing world economy', seems to ignore these hard facts in just the way that the self-deceived agent takes care to hide from his or her own view any information that would be at odds with the story which justifies their present behaviour.

Before we can get to grips with corporate or governmental self-deception, we have to recognize another dimension to social life.

* This appeared in the pamphlet *Sustainable Development: The Business Approach*, issued by International Chamber of Commerce, Paris, November 1989.

† See the discussion in L.R. Brown *et al.* (1990) *State of the World 1990* Unwin, London.

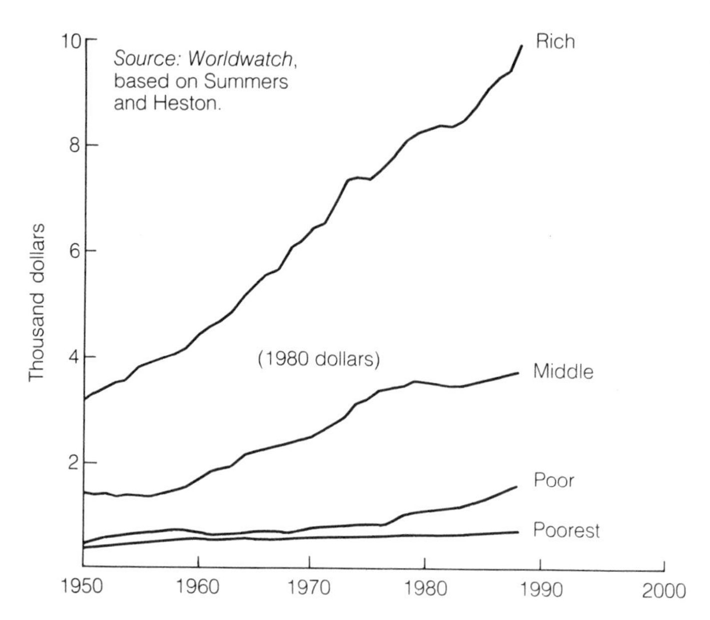

Fig. 1.1 Income per person for four economic classes of countries, 1950–88.

According to Max Weber's methodological individualism, society consists of individual agents ('persons') whose beliefs, decisions and actions are the only determinants of what happens. If he is right, 'corporate actors', as they might be called, have no explanatory role in the account of social happenings. The reduction of sociological explanation to the individual level used to be attractive to those who wanted to model their work on the physical sciences. More recently, however, the growth of systems theory has, among other influences, made space for more holistic forms of explanation within physics itself.

We need not take a final view on individualism versus holism to appreciate that things other than individual human agents have an important role to play in setting the conditions for life in a modern society. Even if the decisions of trade unions, companies, political groups, universities and the churches can in some way be reduced to decisions of individuals within them, there is no doubt that these corporate agents do take decisions*. Moreover, these decisions are often highly significant for the environment. As J.K. Galbraith has pointed out, once we recognize the existence of corporate actors we can also appreciate the difficulty of controlling these agents. Galbraith focused his analysis

* For a Marxist analysis of corporate action, see J.S. Coleman (1982) *The Asymmetric Society*, Syracuse University Press, New York. A general overview of corporate versus individual action is given in B. Hindess (1989) *Political Choice and Social Structure*, Edward Elgar, London. Ch. 4.

primarily on the large transnational corporation but even a small company pursuing growth in a rural community can impact on the lives of its neighbours in ways they may find hard to control*.

Whereas persons are a focus of both agency and experience, larger social agents, like the company, lack experience altogether. As a result, there is no reasoning with them and no room for moral reciprocity of the sort that exists between individual agents. Moreover, the lives of many people can be bound up, directly and indirectly, with the flourishing of a company, a church, university or nation. Think of a factory that produces noxious emissions which local people can both see and smell. Not only can the factory not smell or see its own emissions but its success will typically matter to its own employees, stockholders, other corporate actors (such as the agencies of local government, other companies involved in supplying and servicing, and so on) and even to those consumers whose identity may be bound up with the final product.

In this situation, the factory management, associated companies and the local authority can all regret the nuisance to, or poisoning of, the factory's human neighbours. But, given the company's goals and the ways these synergize with the goals of other individuals and corporate actors, there is no straightforward way of taking account of fairly basic objections to its operations. If my neighbour lights bonfires that interfere with my enjoyment of my garden, I can, in theory, take this up on a person to person basis. But there is nothing to take up, even in theory, with the factory or the company. The lack of shared experience means there is no person-to-factory or person-to-company basis on which to operate. For there is no basis of shared experience on which to establish the kind of reciprocal relationship which is fundamental to many moral dealings.

One solution to this problem, taken in the late nineteenth century alkali acts in the UK, is to use a framework of legislation to protect the individual from the indifference of the corporate actor. The law is able to recognize a category of persons that extends beyond the individual human agent. Such legal persons can be called upon to limit their behaviour or pay compensation and these calls are enforceable. But recognizing the existence of corporate actors and framing legislation to control them is only the beginning, not the end, of the problem. One effect of the British alkali acts was to encourage the building of high chimneys so that pollution could be dispersed beyond the confines of the local area. As the twentieth century progressed and the energy-intensive society flourished, the end result of such policies throughout Europe was the acidification of the lakes, rivers and forests of Scandinavia and Scotland.

* The classic development of this theme is in Galbraith (1967) *The New Industrial State*, Penguin, Middlesex.

1.4 CONSTRAINTS ON DECISION-MAKING

So far, I have argued that there are a number of constraints on environmental decision-making. Some of these only operate at the level of the state: it is for governments and their advisers to be aware of the role of corporate actors in society and to pass legislation that will control their negative effects. Intervention in markets, by way of 'polluter pays' and similar policies can also make it costly for corporate actors to continue to operate in ways that are environmentally destructive. However much control is exercised, there will be none the less problems of consumerism to face as well. It will be a considerable time before social conditions make it easy for us to live in a controlled way within nature.

Awareness of the impact of corporate actors and the tendency to weakness of will can be helpful in other ways. We sometimes use the rhetoric of responsibility when taking risks. Clear examples include official responses to food scares and nuclear accidents. What looks like highly risky and experimental behaviour is often accompanied by a discourse which would be more appropriate to a risk-aversive society. But no risk-aversive society would introduce new chemical agents, disrupt natural systems and develop chemical-intensive agriculture with the speed and incaution characteristic of the industrialized societies. Again, there seems to be more than an element of self-deception involved.

If what has been suggested so far is correct, then it is time for decision-makers and policy analysts to take account of our individual and corporate tendencies to tell ourselves comforting stories that help us live with self-deception and weakness of will. However, awareness of these various factors can provide no more than a background to real decision-taking. Are there more specific steps that we can take when framing environmental legislation, considering action and taking decisions? There are two specific ideals I would venture to recommend, both still at a general level. One is the ideal of *avoiding shallow analysis*, and the other is the ideal of *avoiding widespread technocratic and 'green' myths*. In the present section, I will give some examples of shallow analysis in environmental discussions, leaving myths to section 1.5.

Shallow analysis is a close companion to self-deception. Not looking too deeply into a situation can help make it seem less complicated than it really is. Many examples of analysis that is intelligent, well-informed but at the same time shallow can be found in the recent White Paper produced by the UK Government: *This Common Inheritance*. Of course, it is the nature of government documents to defend the status quo and the government's own record. Likewise, we will expect to find shallow analysis as a feature of various industrial publications (for the same defensive reasons). For my purposes, however, I will be criticizing the White Paper not by political standards, but by pretending that it purports to be a reasoned statement of objectively sensible environmental priorities.

Although the White Paper gives clear statements of the problems surrounding the greenhouse effect, air and water pollution, waste, recycling and land use, it has little to say on what are, globally, the most pressing environmental problems of the day. These include fresh water scarcity; slow response rates (e.g. for ozone depletion, reversing accumulation of pollutants, etc.); loss of biodiversity; transboundary pollution; the role of war, and preparations for it; populations, resources and poverty; and global economic relations and environmental destruction.

There is not total silence on these issues. What is lacking is any proper study of each on its own or a thorough account of its interconnections with the other problems. Thus poverty is mentioned as a cause of environmental destruction but the causes of poverty are not themselves analysed. The loss of tropical forest is deplored but the connections between this and the importation of cheap soya and tapioca feedstuffs for the European bacon industry are not mentioned*.

Tropical forest destruction does not just occur; much occurs in response to economic pressures other than simply the need to service debts. Recent figures show that human beings now use over 40% of the total photosynthetic product of the planet, and this cannot be written off simply to poverty. To change examples, despite rhetoric on transport infrastructure and the benefits of car sharing, there is no recognition in the White Paper that the manufacture and use of the private motor car is a major factor in the depreciation of natural capital, waste of energy and the generation of avoidable pollution.

In fact, the White Paper suggests optimistically that, as yet, there is no need to impose carbon taxes or restrict the use of the private car. Instead, it recommends the extension of car ownership as an extension of democratic freedom and choice. This recommendation is accompanied by two illustrations which themselves merit close study. Here, we might argue, is a case of the self-deceived state justifying its own failures, and doing so in a way that tries to ensure these failures are hidden from view. Likewise, Sartre's self-deceived girl gives in to the seductive overtures of her teacher by pretending not to notice that he has taken her hand during their conversation. She affects such an intellectual interest in the conversation that she – as it seems – does not notice her hand resting in his at all†. So the

* John Carroll of the University of New Hampshire plots the connections among intensive pig farming in the Netherlands, the production of cash crops in the tropics and the loss of rainforest, in a so far unpublished study.

† Sartre writes: 'To leave the hand there is to consent in herself to flirt, to engage herself. To withdraw it is to break the troubled and unstable harmony which gives the hour its charm . . . the young woman leaves her hand there but she does not notice that she is leaving it. She does not notice because it happens by chance that she is at this moment all intellect' *Being and Nothingness* (1958) translated by H.E. Barnes, Methuen, London, pp. 55–6.

White Paper, involved in an apparent concern for democracy and freedom, seems unaware of the cost at which this freedom is to be bought.

Figure 1.2 illustrates this point in a comparison of cars and taxis per 1000 population. Note the parade of national flags and the text: 'Wide car ownership is an important aspect of freedom and choice'. What are we supposed to infer from this diagram? That Portugal is less free than Belgium, or that the Japanese lack choices open to the Italians? Are we supposed to remember facts about the political histories of Portugal, Spain and Japan (the 'bottom' three on the table) that will associate low car ownership with totalitarianism? The figures contribute nothing to our environmental understanding. While car ownership graphed against air

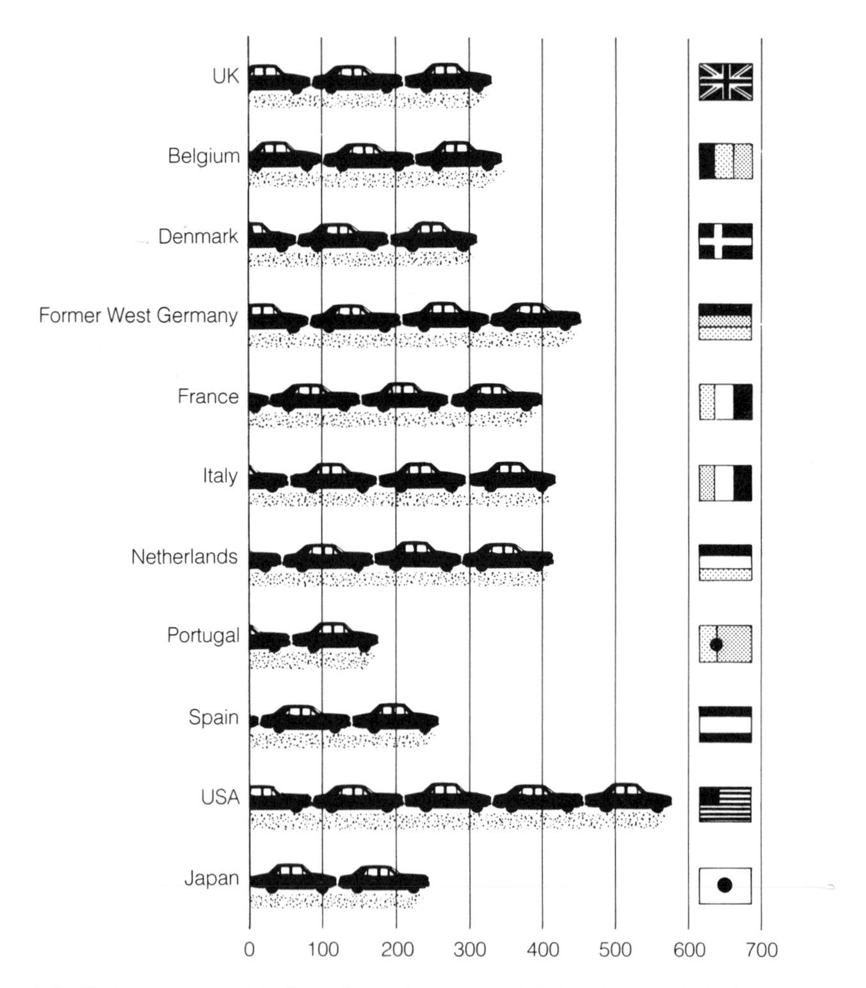

Fig. 1.2 Wide car ownership is an important aspect of freedom and choice (from *This Common Inheritance*, p. 73, by permission).

pollution levels might have yielded some interesting material for reflection (especially in the case of Japan), the presented material is more ideological than informative.

Is the motor car an important aspect of freedom, and 'indispensable for much business travel which in turn is vital for the economy'? The casual reader might not stop to reflect on the implications of these claims. Take freedom of choice, for a start. There are fundamental issues about justice in society which this appeal bypasses. In general, one person's freedom to do something can threaten another's. In a welfarist society, where healthcare is provided by a tax on everyone, the medical costs of dealing with road traffic accidents are borne by all. Air pollution costs are also shared, and include effects on the health of people living in cities and beside busy roads whether they own cars or not. Nor do the owners or operators of cars and other vehicles pay the cost of lost agricultural production or damage to the fabric of buildings. As car ownership increases, the public transport options open to those who lack cars tend to diminish; so the privileges of car ownership lead to a reduction in privileges for those without access to private vehicles. Tax breaks for corporate cars can mean a higher tax burden for the population in general (estimated by Greenpeace to be around £150 per family annually in the UK). Even the freedom of children to play in their local streets or walk in safety to school depends on arrangements for control of motor vehicles.

The costs and denials of freedom consequent on extension of car ownership are separate from the costs affecting drivers themselves. A commuter society need not be one in which time spent commuting is unproductive; trains and buses provide environments in which it is possible to read, study reports or make holiday plans. Driving oneself is dead time, except for the pleasure obtained from driving and viewing the scenery. Finally, those who use their freedom of choice to drive rather than cycle, walk or use public transport, may not be the best judges of their own interest. Driving is not very energetic but often stressful, and hence not the sort of activity that fits naturally into a healthy regime.

Business use of private cars is largely opportunistic. In a society that invests heavily in roads, and does little to control the growth of motor vehicle traffic, it will no doubt be economically advantageous to have sales staff and negotiators on the road a great deal. But there is nothing inevitable about such arrangements. Indeed, there are other technologies which provide conferencing and sales opportunities for businesses – as is well known. If telephones, computers and fax machines have turned the world into a global village, why is it necessary to spend so much time shuttling around it, especially when such shuttling is a major ingredient in the production of greenhouse gases and ground-level ozone pollution? Business travel may one day be seen as an economic disaster rather than as something that is economically indispensible.

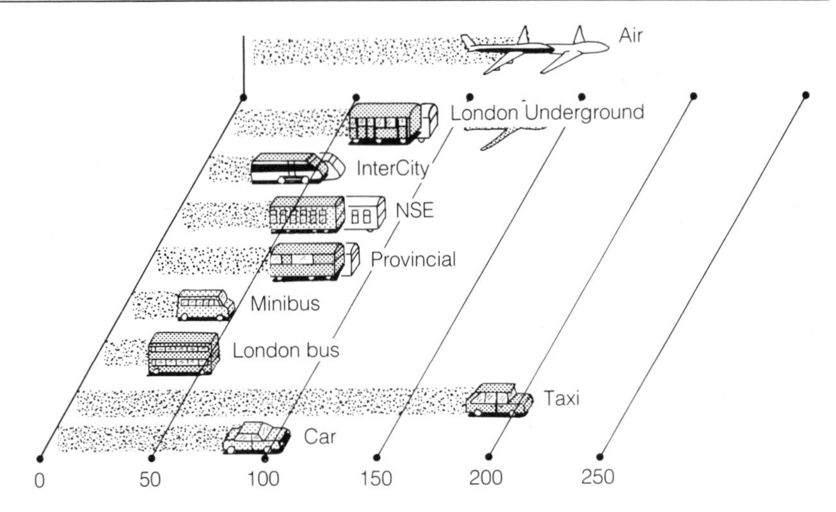

Fig. 1.3 Index of CO_2 as kg of carbon per 100 passenger Km. The figures are merely illustrative of the position in the late 1980s. Technological developments should improve the position of most modes. For example, the Provincial railway should show a significant improvement as new rolling stock is introduced which may roughly halve emissions. Ranges are shown for Air, London Underground, InterCity, NSE, Provincial, Minibus, London bus, Taxi and Car (from *This Common Inheritance*, p. 73, by permission).

Worries about the presentation of transport issues are intensified by a study of the second illustration provided at this section of the White Paper (Fig. 1.3). Here we have an index of CO_2 as kilograms of carbon per 100 passenger kilometres. The choice of this measure indexes private cars at 100, and shows that cars are twice as polluting (per 100 passenger kilometres) as buses, minibuses and Intercity trains. By contrast, taxis are twice as polluting as cars, on the same index, and that is what stands out, given the chosen mode of presentation. But these figures are pretty useless unless we are told just what proportion of the total number of passenger kilometres travelled in the UK each year is related to each form of transport. No such information is given.

The contribution of cars to pollution and the high energy costs of their manufacture and use have prompted some recent studies of the feasibility of large-scale switching to walking, cycling and other environmentally-friendly modes of transport. The widespread introduction of catalytic converters, like the switch to unleaded petrol, simply tampers with the fringes of a problem whose diagnosis and treatment goes right to the heart of industrialized societies*. The motor car is, for many people,

* 'Although exhaust controls such as the catalytic converter have dramatically reduced pollution from US passenger cars since the early sixties, . . . rapid growth in the vehicle fleet and in kilometres travelled have partially offset this progress. Moreover, the catalytic converter . . . slightly increased the CO_2 build-up, which contributes to climate change.' *State of the World*, 1990, p. 123.

undoubtedly the most significant invention of the past 200 years, and the one piece of technology they would least like to abandon. A recent newspaper article claimed that 1 year's losses of oil from the burning of the Kuwait oil wells in the aftermath of the Gulf War could be compensated for in a single year by improving the average fuel efficiency of the North American car by 30%. Significantly, during the 1980s, the average fuel efficiency of cars in the USA declined slightly, even though we now possess the technology to make improvements in economy without noticeable loss in 'performance'.

Clearly a great deal more could be said on transport issues, on the question of how business organizes itself and on the issue of freedom of movement by car. Any serious attempt to tackle both local and global energy and pollution problems would need to explore this area in some depth. The failure of the White Paper to engage in anything other than merely shallow analysis, suggests that the UK Government at the time had yet to shake off self-deception in discussions of these topics. Environmental decisions based on shallow analysis, on hoping that things are not so bad as feared, are not the decisions of rational beings confronting their situation with honesty. It was one of the weaknesses of Greek philosophy, and one that has largely persisted in Western philosophy up to the present, to describe the human situation as one in which purely rational beings are confronted with choices over their future. We are certainly rational on occasion; but our rationality is liable to be clouded by other factors, and to be enlisted in the service of our attempts to hide unpleasant aspects of our situation from ourselves.

1.5 MYTHS AND THEIR AVOIDANCE

If shallow analysis is one trap for the unwary planner, the failure to make due allowance for widespread myths is another. In the recent literature on economics and environmental policy there has been an impressive debunking of the claims of cost-benefit analysis to provide a rational basis for making policy decisions. I suspect that risk-benefit analysis will also fall under the impact of similar attacks, so I will not enter into any specific discussion of these issues*.

Economic theory is not the only producer of myth in this context. I will

* For well-informed criticisms of cost-benefit analysis and associated techniques (such as contingent valuation) see Holmes Rolston III (1985) Valuing wildlands, *Environmental Ethics* 7, reprinted in *Philosophy Gone Wild*, Prometheus, Buffalo, New York, 1986; B. Norton (1987) *Why Preserve Natural Variety?* Princeton University Press, Princeton; and M. Sagoff (1990) *The Economy of the Earth*, Cambridge University Press, Cambridge. For an attack on the Pearce programme see Adams, J.G.U. (1990) Unsustainable economics. *International Environmental Affairs*, **2**. For a more general attack on economic rationality see Brennan A. (1992) Moral pluralism and the environment. *Environmental Values* **1**.

indicate briefly two very widespread myths that seem to me to receive insufficient notice. Interestingly, some writers on environmental ethics and policy seem to believe both of them, while others adopt one of them as part of rejecting the other. One of them can be labelled the *myth of restoring nature*, while the other is the *myth of natural diversity*. Whereas the first myth is meant to underwrite our interventions in nature, the second is supposed to provide a foundation for a policy of non-meddling and non-interference in natural systems. The former appeals to those who suffer from technocratic optimism, while the latter is often endorsed by those who are 'deep green'. Neither myth is wholly false but, as always, discovering the correct degree of credence to place in each is by no means easy.

It has become fairly commonplace in the environmental literature to write off projects for restoring nature to some former condition. Not only can we try to bring about such restoration after open-cast mining and other temporary disruptions to rural areas, but we can try to compensate for destruction of one environment by carrying out a recovery project in another (e.g. by 'moving' a wild flower meadow in advance of developing a green field site). Paul Taylor, in an influential book, has even suggested that it may be our duty to improve various environments by way of making amends for our past treatment of nature*. This is an idea with lots of appeal. For example, it is hard to see how industrialized countries can have the moral authority to urge poor countries to save their forests when the former have already destroyed much of their own indigenous woodland. Moreover, when new planting is undertaken, this may involve fast-growing, high-input non-native species planted for profit rather than with sensitivity to the landscape. Until recently, this was the situation in Scotland. Perhaps things would be different, however, if the forest of Caledon were to be restored and wolves were reintroduced as natural controls on the red deer. If we made the effort to do these things in Scotland, would the UK then have more authority to give advice to other countries about how they should manage their own environments?

Some time ago, the whole project of restoring nature came under philosophical scrutiny in a paper by Robert K. Elliot†. His argument was extremely simple. Given the importance of its causal history, and the particular circumstances surrounding its production, of a work of art, we accord less value to a fake than to an original work. But each landscape or ecosystem is itself the product of historically particular forces working over time to produce its unique blend of populations. The human recreation of this after its destruction is, at best, a forgery and not the

* This is discussed in Taylor, P. (1984) *Respect for Nature*, Princeton University Press, Princeton.
† This paper 'Faking Nature' was published in 1982 in *Inquiry* **25**.

return of the natural landscape or system. Moreover, Elliot argued, the notion that it is even physically possible to restore a lost ecosystem reveals a 'technological mind set', as if nature is something that can be tamed, harnessed and manipulated for human purposes. If Elliot is right, then there can be no prospect of restoring the forest of Caledon.

Even the most enlightened can fall foul of Elliot's criticisms. Chris Maser, a forester disillusioned with the insensitivity of the USA forest service, published in 1988 a text entitled *The Redesigned Forest*[*]. In it he argues that nature has a discoverable design, but that human beings lack both the parts catalogue and the maintenance manual for building a forest. As a result, we must stop destruction of the remaining old-growth forest, mature forests and young-growth forests, so that we can have 'a parts catalogue, maintenance manual and service department from which to learn to practice restoration forestry'. An additional benefit of this policy, as Maser points out, is that restoration forestry itself will meet myriad human needs, not only those requiring the production of fast-grown wood fibre.

As Eric Katz has recently pointed out, the language and attitudes of Maser's treatment are highly technocratic[†]. Moreover, it looks as if restoration forestry might fall foul of Elliot's criticism. However well we mimic natural forests, surely the best we can hope for is a fake, something lacking the historic particularity of the natural forest. Even taking steps now to replant the old-growth forests of British Columbia will not, according to Elliot, give us something of value equal to what has been lost. But, however inferior, perhaps a forgery is better than no compensation at all. Katz argues, though, that the issue of fakery is not the central one. What matters is the attitude of domination, an attitude which restorers as well as destroyers can share, and the fact that the restored woodland will be a product of human manipulation and interference. For Katz, what is most natural is most free from human interference: a restored nature is a fraud, not because it is a fake, but because it is an artefact.

I will leave aside here many of the issues at stake between Elliot and Katz. Neither, it seems to me, has compelling analogies to offer. A forged work of art is just as much a human creation as its original; thus Katz is right to wonder whether the analogy with art is of help when we are considering the value of a restored nature. But Katz himself is silent on the definition of what makes an artefact. Is a garden an artefact, just as a car is? What about a wild garden? Or a neglected garden? Are national parks and game reserves artefactual to a lesser degree than commercial

[*] This was published in 1988 by R. & E. Miles, San Pedro.
[†] His points were put in a paper, Katz, E. (1992) The big lie: human restoration of nature. *Research in Philosophy and Technology*, **12**.

plantations of sitka spruce? On these questions, Katz is silent. In an age of genetically-engineered organisms, the issue of the status of artefacts versus natural things is liable to be a vexed one. Despite these problems, it is clear that the issue of whether and to what extent nature can be restored is a difficult one. There is no easy formula to put forward here. Moreover, restoring lost species is something that no one would claim to be able to do at present, and the current rate of species elimination in the tropics makes any talk of future restoration look like pure wishful thinking.

Although Katz is too sophisticated to be caught by this trap, we have to be careful of falling into the belief that what is free from human interference is good in some ultimate, 'wild' way*. The notion of the goodness of wild nature is present in many romantic thinkers, and has been a prominent feature in the growth of American environmental awareness. Amidst the complex of beliefs about the goodness of natural systems is the other myth that I want to mention, namely that nature is naturally diverse. For Aldo Leopold, the prime ecological value was the good of things 'natural, wild and free', and ecosystems, left to their own devices showed 'beauty, integrity and stability'†. The notion that nature left to its own devices will produce landscapes of great beauty comes over clearly in the rapturous prose of the early white settlers of the San Francisco and Monterey Bay areas. Finding this area to be idyllic, covered in open meadows and park-like forests, they thought they had chanced upon a natural environment of sublime enchantment. They were wrong. The Ohlone Indians had been managing the Bay area for a very long time, subjecting it to periodic burns to prevent the predominance of dense woodland‡.

Elsewhere, I have described the studies of H.S. Horn on New England woodlands. Horn's researches show that such forests have a clear tendency away from diversity, towards domination by a few species§ Periodic fires, and the impact of humans by way of clearing portions of forest, maintain what diversity there is. There is little tendency to balance or diversity in such areas. Indeed, the whole issue of whether ecosystems are generally self-maintaining diverse systems or simply fortuitous groupings of populations that at least for a time are not lethal for one another is very much undecided in ecology. It is striking, and unfortunate, that many conservationists operate with ideas of balance and

* For a recent, impassioned statement of the anti-interventionist position see McKibben, B. (1990) *The End of Nature*, Viking, New York.

† This is explored in Leopold, A. (1949) *A Sand County Almanac* Oxford University Press, Oxford.

‡ See Margolin, M. (1978) *The Ohlone Way* Heyday Books, Berkeley. Margolin is quoted in Bookchin, M. (1990) Recovering evolution. *Environmental Ethics* **12**.

§ Horn, H.S. (1975) 'Markovian Properties of Forest Succession', in Cody, M.L. and Diamond J.M. (eds) *Ecology and Evolution of Communities*, Harvard University Press, Harvard. For a simplified account of the tendency to uniformity see Brennan, A. (1988) *Thinking About Nature*, Routledge, London, Ch. 7.3.

diversity in nature that were more prevalent in the nineteenth century than among contemporary ecologists.

What we may regard as a pleasingly diverse environment may be one that requires active management on our part. Moreover, there is little point, I would suggest, in arguing that nature should be left free from our meddling. Non-interference in nature is no more an option for us than for any other mammal. It is the form of our meddling which raises moral, aesthetic and policy issues. In saying this, I am not arguing that there are no cases for land to be put aside or protected from human interference. There are many reasons why the Antarctic, for example, should be set aside as a wilderness park, safe from mining and, indeed, protected from overmuch scientific investigation. At a more local level, there are countries where there is no wilderness. What is sometimes thought of as wilderness (e.g. much of the Scottish Highlands) may well be an area created by human intervention and maintained by such intervention. But there are countries where there are areas that are relatively wild (as some small parts of Scotland, and larger parts of the USA and Australia). I would argue that the treatment of such areas is a matter of concern and, again for a number of reasons, at least some of them should be left relatively free from human meddling.

The really difficult and interesting issues arise, it seems to me, for areas that are clearly home to groups of people, and where the recent impact of human populations has led to species loss, erosion, changes in the water table and so on. As already pointed out, some of the forces driving these impacts will probably originate far from the places where they occur. But in such cases non-intervention is not an issue. What is at stake is the quality of life of the people and other species involved, and the ability of human beings to pass on to their children lands that are in as good, or better, condition. We cannot trust optimistically to technology to fix our problems if a change in lifestyle is required so that we can be gentler on our surroundings. Nor can we hope that nature left to herself will know best when we are already heavily committed to changing the nature of our surroundings for better or for worse (and generally, it has to be said, for the worse).

1.6 CONCLUSIONS

The argument of this chapter has been conducted at a somewhat general level. What I have tried to do is establish a number of points that are not widely recognized in the literature on environmental policy. A sustaining belief behind the comments here is that until we do get to grips with the points identified here we will fail to take the real measure of the environmental problems that we face.

Initially, I claimed that a widespread but puzzling phenomenon underlies a great deal of our inability to take the proper steps to tackle environmental issues. This is the plight of those who are *akratic* or 'incontinent' in Aristotle's sense. We lack the control to take the right course of action even when we know what it is and desire to take it. In more recent philosophy, the recognition of self-deception or 'bad faith' has given us means to articulate, even if not to cure, another widespread human failing.

Self-deception and incontinence have attracted philosophical interest not just because they are widespread but because they are also deeply puzzling. If we really desire to pass on the earth in good shape to our children, then why do we not act on this desire? If we really want to understand our situation, then why do we not give detailed attention to our way of life and our economic situation instead of opting for shallow thinking and myth? It is as if in the one case we desire something, yet do not desire it, and in the other know something but also seem not to know it. The existence of these problematic conditions can be a source of pessimism. Sartre's bleak version of existentialism is deeply involved with his diagnosis of our condition as one of bad faith and an associated failure to recognize our true freedom. But on a more optimistic note, we might argue that to implement environmental policies successfully we will need to take account of the general tendency for both individual and corporate self-deception. In this case, forewarned is, to some extent, forearmed.

It is neither unfair nor unkind to governments, public agencies and corporations to observe that we are a long way from full honesty in our debates and deliberations on the environment. Some of the reasons for the lack of honesty may be inescapable features of our, i.e. the human, situation (as Sartre believed). But not all of them are. Even if we do slide into bad faith from time to time, we do not need to assist this by acceptance of shallow analysis and mythic portrayals of our situation. Nor need departures from virtue through weakness of will be encouraged by rationalizing our situation in the ways illustrated by the examples from the White Paper on the environment. On the contrary, to make progress in tackling our increasingly desperate environmental plight we have to make strenuous efforts to overcome our myth-making and ready acceptance of partial, shallow versions of the truth.

The examples I have given support two ideas. One is that shallow analysis and myth is present in the literature on environmental policy. The other, obvious from the very disclosure of these failings, is that these particular failings can be overcome even by creatures who themselves are perhaps prone to other, more deep-rooted failings. It is particularly unfortunate, when we consider the mythology of the conservation movement, that many of those who regard themselves as morally 'deep' fall foul of the accusation of shallowness. Their tendency to adopt the simplicity of a comforting set of myths should not, of course, blind us to what is worthwhile in the moral position they adopt. But to assess the merits of their moral position is a task for another occasion.

The net result is thus a blend of pessimism and qualified optimism about our present and future environmental decisions. I have accepted for the most part that we are always going to be prone to self-deception and incontinence. These may well be deep-seated and largely unchangeable aspects of our lives, but there is nothing equally inevitable about shallow and mythic thinking. These failings can be identified now and guarded against both for the present and the future. If I had to identify two forces in our intellectual lives that can help with the latter task they would be good science on the one hand and good philosophy on the other. As long as both of these are cultivated there are at least qualified grounds for optimism; but their cultivation alone will not bring about change. To make any impact, they need to be adopted and put into practice by those involved in policy discussions, corporate planning and environmental activism. That environmental decisions should be informed by good science and philosophy may seem hardly a controversial conclusion. Our route to it, however, suggests that putting such an innocuous resolution into effect may be far from easy.

Environmental attitudes in North America 2

Frank B. Golley

Frank Golley is Research Professor of Ecology at the University of Georgia, in Athens, Georgia, USA. He was President of the International Association for Ecology (INTECOL) 1986–90.

2.1 INTRODUCTION

North America has been an arena for a variety of environmental dilemmas which, while not unique, represent different problems and responses to those in Europe, Africa or Asia. North America was essentially unknown to Europeans until the end of the fifteenth century and settlement occurred relatively slowly. The Spanish explored the southern and western sections and occupied the southwest and California. The most successful settlements were by largely disaffected groups from the British Isles, with additional contingents from Germany and France and from Africa. North America has continued to receive immigrants from all parts of the world.

The initial dilemma caused by this process concerned the interaction of Europeans with the existing inhabitants of North America. After all, the continent was fully inhabited by a rich assortment of native people who practised a variety of technologies. New settlers were faced with occupying other people's land, with developing ways to interact with the existing inhabitants and learning how to survive under new environmental conditions. This effort of adaptation was very important in the first years of settlement, but became less important as the self-confidence of the European settlers increased and in due course became replaced by a contempt, hatred and purposeful aggression against native Americans.

Environmental Dilemmas Ethics and decisions
Edited by R.J. Berry
Published in 1992 by Chapman & Hall, London. ISBN 0 412 39800 1

We will consider a case example of this phenomenon as our first environmental dilemma.

Settlement not only involved interaction with the inhabitants and owners of the land but use of the natural resources to establish European patterns of life in the New World. The continent was rich in natural resources which could be exploited by new technologies. The natural resource which was of special importance at the beginning of settlement was the forest (Lillard, 1947). Forests provided the wood for Britain's navy, especially when access to Baltic wood was prevented, as during the first Dutch War of 1652. The King's foresters marked great pine trees for ship masts and bent live oaks for timbers of naval vessels; special ships were constructed to take these resources to England. The forests provided much of the material and energy for the emerging American industrial revolution, and the lake states' forests and western forests provided the lumber for expansion of settlement. Overcutting, exploitation without a reforestation plan, destruction of streams, and great wild fires were all associated with the lumber industry. Public concern at the turn of the nineteenth century led to the creation of the US Forest Service by President Theodore Roosevelt in 1905, with Gifford Pinchot as chief forester. Pinchot was a utilitarian conservationist who believed that wise use of forests, not preservation, was the best policy (Pinkett, 1970). Conservationists, such as John Muir, disagreed with Pinchot's policies (Fox, 1981). The conflict over forest management continues today and represents our second case study of an environmental dilemma.

European explorers and colonists encountered a continent that was immense, seemingly inexhaustible, and dangerous. It required hard work, technical skill and luck to succeed in the wilderness. Settlers replicated successfully European lifestyles and social organization, especially in New England, in Virginia tidewater areas and in areas settled by Germans from the Rhine Palatinate. However, the American landscape, after the revolution, gave birth to a new individual, the frontiersman or pioneer. The frontiersman became the mythical hero who opened new lands to settlement through his courage, skill and derring do. The frontiersman and his surrogates, the mountainman, the cowboy, the logger, the flatboat man, represented in an exaggerated way the characteristics Americans considered valuable and desirable.

In many cases, frontiersmen were from Ireland and Scotland on the Celtic fringe of Europe. The cultural characteristics of Celtic people were adaptive to the American frontier. Indeed, their life style and appearance was sufficiently similar to the American Indian that the Indian was interpreted to a popular audience in England, at the time of settlement, by reference to an Irishman (Quinn, 1966). Not only was personal conduct and lifestyle adaptive, the typical American crossroad with church, post office, store and mill was derived from the central settlement of Scottish settlers in Northern Ireland (Evan, 1969). Thus, the Celtic character adapted the settlers from Scotland and Ireland to the conditions of the frontier. Other Americans imitated them.

At the end of the nineteenth century Frederick Jackson Turner, a major figure in American historical scholarship, declared that the frontier was no more (Turner, 1893), although pressure to continue the advance persisted and became international in the Spanish American War of 1898 and various invasions of Mexico and Central America. However, the frontier as a testing ground for character had not truly disappeared. National parks, forests and later wilderness areas were established where young people, especially, could encounter the challenges of nature. Wilderness was a concept created largely by Americans to refer to those natural areas which provide experience of the conditions that were thought to have moulded the American character (Nash, 1982).

The environmental dilemma comes from the conflict between the wilderness advocates who want to preserve nature as it was when the frontier existed and the users who represent other frontier types, such as the logger, cowboy and miner, and see the wilderness as an economic resource. An additional problem is that to maintain a landscape in a setting characteristic of the past requires management. Nature is changing, and human demand, perception and knowledge is changing. How does one manage a local landscape to maintain constancy under such dynamic conditions? While wilderness is quintessentially an American concept, it illustrates a dilemma that is universal.

The purpose of this chapter is to illustrate some environmental dilemmas in North America, especially in the USA, and to use the case studies to show how Americans face environmental issues, especially from an ethical perspective. Of course, in a brief chapter I cannot do more than suggest the nature of these complex issues. North America is a large area and North American society is complex. Response to environmental issues is always multifaceted and Congress, state legislatures, judges and citizens always have to weigh various interests for and against any issue. My choices are intended to represent what I feel are some broad historical trends of general significance.

2.2 ENCOUNTER WITH NATIVE AMERICANS

At the time Europeans first visited North America as explorers and settlers, the native American Indian populations were a vigorous and culturally dynamic people distributed across all environments of the continent; their cultures were finely adapted to the resources of the environments they inhabited, and where possible these people created cities, fortresses, apartment houses, complex governments and other characteristics of so-called advanced civilizations. Unfortunately for them, American Indians were not resistant to many European diseases and their numbers were decimated and their social systems were destroyed by death even before face-to-face encounters with Europeans occurred (Crosby, 1986). Diseases swept before the Europeans, weakening native people and making them unable to resist the invasion of their homelands.

Nevertheless, the initial encounters between Indians and Europeans were usually peaceful, and the Indians were eager to trade and helped the Europeans adapt to new environments (Cronan, 1983). However, the two cultures could not co-exist. The newcomers were obsessed with land as property and hence with the boundaries of property; Indians perceived land as a communal environment where boundaries fluctuated with season and habitat. While Indians quickly adopted European technology, they seldom adjusted to the idea that the environment could be divided up by individuals and treated in any way an individual desired. Even today, the reservations are considered the common property of the tribe, where individuals have rights to use resources, but only in the context of the whole tribe.

The European was puzzled by the Indian. The Spanish debated whether Indians were animals or men. Various Europeans, such as Samuel Penn, the Quaker, attempted to understand the Indians and treat them honourably, in the way that Europeans conceived honour. However, the vanguard of Europeans who penetrated Indian territory to trap beaver, graze cattle and hunt were frequently of Celtic origin, highly adapted to frontier life, violent and ruthless in defence of their self-defined perogatives, and skilled in the extensive use of natural resources (McWhiney, 1988). The frontiersman became a mythical hero and after the American revolution, when the government came under the influence of frontiersmen, the relation with the Indians was translated from one of legal co-existence into one of annihilation.

An example of this is the fate of the Cherokees who lived in the southern Appalachian mountains and Piedmont region of Georgia, North and South Carolina and Tennessee. These people, through contact with missionaries and with government assistance, had become sophisticated in a European sense, with a written language, a newspaper, living in wooden houses and farming lands, some even with African slaves (Vipperman, 1978, 1989). Gold was discovered in Cherokee land in 1829 and the US Government signed a fraudulent treaty with a renegade, unofficial group of Cherokees which agreed to move the Cherokees to Oklahoma, paying them five million dollars for their land and improvements in the southern Appalachians, and giving them land west of the

Mississippi River. The leaders of the Cherokee nation protested against this false treaty but the US Government sent troops to enforce it. The transfer was to be accomplished over 2 years. Payment and transfer was to be directed by a commission headed by the governors of Georgia and Tennessee. The Governor of Tennessee, William Carroll, never participated, thereby delaying the payments of Indian claims under the treaty for over 6 months. After a new commissioner was appointed, the commission had gold coin for only some of the payments. Obviously, this bungled operation of the Federal government dictated that the treaty dates be extended, but the Georgia governor was adamant that Cherokee territory be open to any settler on the date of the treaty, thereby preventing further delay and adjudication of claims. As the date drew near, troops rounded up Indians and began a march in the dead of winter to Oklahoma. When they reached the west in March 1839, over 4000 Indians had died of starvation and violence (Vipperman, 1982). This is called the 'Trail of Tears', and it is an example of a recurrent violation of human rights of the American Indians (Williams, 1990). Even today, the US Supreme Court has ruled that communal development of natural resources on reservations is unlawful, thereby preventing them from functioning as Indians. Even so, when reservations are large enough to contain sufficient resources, Indian cultures have survived. As Indians have become educated in legal matters so that they can defend themselves in the alien culture, they have slowly exerted their claim to existence as tribes, and currently there is a renaissance of pride among native Americans and a growing respect by American non-Indian citizens for Indian rights.

The ethical dilemma is this: how can we accept human diversity in our environment, when the other individuals view the environment in a fundamentally different way than we do? How does one find a compromise with people who do not divide the world into property and who do not convert nature into a resource that can be exploited to extinction on a theory of economic convertibility? To do so requires a self-confidence and self-knowledge that is frequently missing in contemporary Euro-American culture. I see no simple answer to this question. But it requires an answer, especially as the world grows more crowded and we literally rub shoulders with people who do not share our environmental and social perceptions.

2.3 ENCOUNTERING THE GREAT FOREST

The forest encountered by European settlers in America was incredible. The words of John Bartram, a botanical explorer of Georgia in 1773, illustrate what one encountered in many places and can still see in a few

National Parks (van Doren, 1928). In the following quotation Bartram is referring to a landscape near Augusta, Georgia:

> Leaving the pleasant town of Wrightsborough, we continued eight or nine miles through a fertile plain and high forest, to the north branch of Little River, being the largest of the two, crossing which, we entered an extensive fertile plain, bordering on the river, and shaded by trees of vast growth, which at once spoke its fertility. Continuing some time through these shady groves, the scene opens, and discloses to view the most magnificent forest I had ever seen. We rose gradually a sloping bank of twenty or thirty feet elevation, and immediately entered this sublime forest. The ground is perfectly a level green plain, thinly planted by nature with the most stately forest trees, such as the gigantic black oak (*Q. tinctoria*), *Liriodendron, Juglans nigra, Platanus, Juglans exaltata, Fagus sylvatica, Ulmus sylvatica, Liquidambar styraciflua*, whose mighty trunks, seemingly of an equal height appeared like superb columns. To keep within the bounds of truth and reality, in describing the magnitude and grandeur of these trees, would, I fear, fail of credibility; yet, I think I can assert, that many of the black oaks measured eight, nine, ten, and eleven feet diameter five feet above the ground, as we measured several that were above thirty feet girth, and from hence they ascend perfectly straight, with a gradual taper, forty or fifty feet to the limbs; but below five or six feet, these trunks would measure a third more in circumference, on account of the projecting jambs, or supports, which are more or less, according to the number of horizontal roots that they arise from: the tulip tree, liquidambar, and beech, were equally stately.

However, these great forests were frightening to Europeans. In their letters and diaries they frequently remarked how wonderful it was to remove enough trees to see the stars or, even better, to see a neighbour and know that you are in a humanized landscape (Lillard, 1947). The forests were so vast and so large that few thought that they could ever be cut down. A peculiar type of male labourer, the logger, evolved and the mythical figure of the giant logger, Paul Bunyon, came to represent the heroic battle of man against the forest (Lillard, 1947). Stories of Paul Bunyon and his giant blue axe are still popular subjects of folk tales and songs in the Lake Region and the Pacific Northwest.

However, the forests were felled and, in some regions, so many trees were removed that there was not enough wood left to cook food, so that people had to move to find fuel wood. In other areas the wastes left on the land caught fire. In Wisconsin and Michigan these fires became extensive conflagrations which burned forests, and caused large loss of human life. The Peshtigo, Wisconsin fire of 1871 was one of the worst, burning a million and a quarter acres and killing over 1100 people (Lillard, 1942). Eventually, the public recognized the environmental damage caused by logging and supported the organization of a US Forest Service, which became responsible for the forest and range land still under government jurisdiction. Even so, the Forest Service was faced with many claims on the use of forest land, and it could not adopt policy that was unacceptable locally.

The environmental dilemma in this instance involves management with multiple conflicting objectives. The opposing user groups (loggers, hunters, recreationists, conservationists) have few basic ideas or needs in common. Frequently they do not even live in the same region. The resources available in the groups to manipulate public support differ widely, and the Forest Service often has had to take a position contrary to the goals and needs of specific user groups. Where there are consistent patterns of the Service supporting the goals of one group over long periods of time, such as logging interests, other interests become frustrated and can result in violence. For example, Earth First has recently emerged as a force for protection of the environment in the western USA (Scarce, 1990). Earth First members drive metal spikes in trees to prevent them from being sawed in mills, sabotage logging equipment, and remove survey stakes for roads and development; they justify this economic violence as an ethical activity. While the number of Earth First members is probably very small, there is widespread support and approval of their actions among the public who perceive loggers as people who have had special help from the taxpayer through access to trees planted and managed by public funds and who want to destroy old-growth forest for private gain.

The environmental dilemma illustrated by this case study is the problem of satisfying multiple interests in a social and political environment where there has been no history or mechanism of compromise. In countries such as Sweden, where most citizens share environmental attitudes and offical commissions involve all the parties of interest, compromise positions can be developed and frustration is reduced among interest groups. However, in the USA the solutions are obtained through an adversarial relationship that pits one side against the other, often in a court of law. This approach to environmental disagreement fits the penchant of American society to cast interaction in competitive, economic terms. The presence of abundant resources on the frontier has permitted the society to avoid the most serious costs of the approach. Frequently, the costs of competition are paid by the environment or passed on to future generations. With abundant resources the losers could move to another place and try again. However, as resources become limited through population growth, over-use and changed demands, pressures on them increase and there is now growing frustration as larger and larger segments of the society are denied opportunities. A new approach to resource allocation is needed in American society.

In this situation, the land ethic of Aldo Leopold has been a steady compass for many environmentalists. Leopold declared that actions that preserved the integrity, stability and beauty of natural communities were ethical and good (Leopold, 1949). Actions that did otherwise were unethical and bad. While Leopold's emphasis on maintaining natural order is influenced by deterministic ecological concepts, such as the succession and climax theories of Frederic Clements, which were widespread at the time he was active, his ethos provides useful guidance in the resolution of this environmental dilemma.

2.4 WILDERNESS AND NATIONAL PARKS

My final case study involves a particularly American obsession: the concept of wilderness. Wilderness represents an extensive area where human development is absent and where natural forces are allowed to operated uncontrolled. Wilderness represents in an entirely mythical form the American continent before European settlement. It ignores the activity of the American Indian, the fact that wilderness areas are tightly coupled to larger landscape regions, and the fact that the environment is continually changing. However, it does provide an opportunity to experience nature in a way that is otherwise impossible and represents a uniquely unmanipulated landscape where humans can escape humans for a moment.

As the American landscape was brought under private ownership or governmental management, the public recognized the need to set aside certain areas of great natural beauty which had no obvious economic value. Yellowstone National Park was among the first of these reserves, and the political debate over its formation was intense (Chase, 1986). Nevertheless, public pleasure in the preservation of Yellowstone, and the growing number of other parks amply justified a policy of park expansion. However, here too the National Park supervisors had problems of management for multiple objectives, and shortly after World War II, a commission headed by A. Starker Leopold of California, the son of Aldo Leopold, examined this problem and recommended that the parks should be managed for those patterns of wildlife, vegetation, and ecological condition that existed when the parks were first visited by Europeans (Chase, 1986).

The consequences of this decision were enormous. Vegetation and wildlife managers attempted to return the ecological condition to that described in the diaries or journals of explorers, interpreted by ecologists and historians. Natural processes were allowed to operate only so far as they recreated earlier conditions. Human use of parks and wilderness were managed to fit these objectives. However, many of the park environments were exceedingly fragile, many tended to fluctuate widely, and the goals conflicted more and more with the ecologists' understanding of a dynamic, non-static, non-equilibrium, natural system. Park management policy began to change, allowing natural processes to operate uncontrolled. However, this more modern form of management resulted in conflict with public and private demands. Allowing mountain goats to die of disease in Glacier National Park, and wildfire to burn uncontrolled in Yellowstone National Park, resulted in management crises. The response was to bureaucratize management. In Yellowstone, rangers are required to file daily fire reports, and the supervisors must guarantee that they have the resources to contain a fire. The process of park management, while always political, has become so politicized as to destroy the enthusiasm of the employees and drive able people to other professions. Politicians respond to local public pressure based on partial and often misinformed opinion, and their compromise benefits no one but their own re-election.

The environmental dilemma created in this case study involves the development of management objectives that represent our best knowledge of how nature functions in a public arena which has multiple, conflicting private interests, imperfect knowledge of the system, and is tied to the political process. Ironically, a policy of managing for a mythical past condition fits public interest and an outdated deterministic ecology better than does a more modern policy, which would allow natural processes to operate within broad limits. The ethical problem of setting aside land as wilderness, which only the physically fit can enjoy, in a form which ignores natural and social reality, is difficult to resolve. How far should a park be developed to provide universal access and urbanized entertainment? How can public education teach the history of the country and natural science so that citizens can express their private views within realistic boundaries? How can local needs and national demands be harmonized?

2.5 CONCLUSIONS

In this chapter I have discussed three case studies which represent environmental dilemmas in North America, but which appear to have a broader significance. First, I have asked how can we manage natural resources as a resource for human use in a world where human cultures have widely different ways of perceiving nature? All nations are linked together by the atmosphere, by ocean currents, and by the migration of animals, plants and microorganisms, including disease organisms. All nations share common resources. Yet, we have no global management philosophy, nor do we have an international system that would allow us to implement such a philosophy.

The United Nations represents a forum where we can discuss these vital issues, but success in global environmental management has been mixed. Powerful nations, able to exploit the global resources or which pollute the common environment, tend to create compromises through treaties which are beneficial to their economic interests. Failing development of a compromise, they resort to economic or military aggression to satisfy their demands. Clearly, this *ad hoc* system needs to be replaced if we are to manage global environmental change.

The second case raises a different set of problems. Here the question involves the policy for management of a landscape for multiple objectives. In this situation there are truly winners and losers as you cannot clearcut a forest and preserve its natural beauty and recreational value. How can we find a compromise between multiple demands where human livelihood and economic well-being is at stake? The question is universal, but is especially acute in a democracy where all citizens have access to public resources and express these rights through the media, political process and the courts. American respect for law has led them to use the legal process as the ultimate arbiter in this dilemma. If the legal process is unfair or too slow, frustration may find an outlet in violence and sabotage.

Finally, on the most local scale, how do we manage an environmental resource in an environment where our scientific understanding of natural processes is changing rapidly but public understanding is changing much more slowly. Scientists and technicians give different answers to the same questions at different times. The intellectual environment changes quickly. Nature itself is also changing, especially under the impact of human disturbance. Popular culture and the political system change more slowly. The consequences seem to be perpetual conflict. In America the debate about environmental ethics tends to be carried out within the media, such as television and news magazines. The public then responds to the issues as defined by the media through organizations, such as the Sierra Club, that influence the decision-makers, who must resolve the issue. Because conditions are continually changing, the compromise of the moment eventually unravels and the debate continues. In this way public action follows and is driven by the environmental crises. The task is to create plans that anticipate and guide change.

In my judgement, these issues can be resolved best through a common respect for and appreciation of factual knowledge about the environment and a common environmental ethic. While scientific information and concepts are culturally and socially controlled, the physical, chemical and biological features of the biosphere are sufficiently well known that we can reason from them with a specified degree of certainty. We need to be careful about extrapolation from factual data and mixing hypotheses, opinion, theory and law. But, even so, our recognition of the planet Earth as a single object in space, which can be treated as composed of an atmosphere, hydrosphere, lithosphere and biosphere, provide a universal ground for global environmental management. To treat the factual basis of physics, chemistry and biology as mythical and subjective deprives us of the common material with which to think about the environment and act rationally within it.

In addition to this factual base, we also require commonly-held ethical views regarding the environment. I personally feel that this common ethical system should consider both our personal capacity to know and relate to other people, other living beings, as well as to the physical, inanimate environment. Arne Naess (1973), the environment philosopher, terms this capacity 'self-realization'. Self-realization is grounded in the ecological concept of relationship (Golley, 1987). Naess stresses the need to recognize implicitly that we are ecological beings who are open to the environment continuously.

The second foundational concept for a common environmental ethic concerns the way we act as we encounter our environment through self-realization. Naess believes it is essential that we accept other living beings and the physical environment as equal to humans in an ultimate sense. This means that we do not impose a hierarchical order on the world, which implies that humans or some humans have dominion over other beings and the environment. Of course, we expect to encounter enemies, parasites, predators and diseases in our interaction with the environment and these elicit a negative response if they attack us or our factual knowledge leads us to expect imminent danger. Otherwise we 'love our neighbours' until our neighbours prove false. This norm has been called 'biocentric equality' by Naess. I interpret the norm of biocentric equality to mean that we should act in the world as if it were our home and we are among our brothers and sisters. However, I would extend the family metaphor beyond humans to all life and the physical environment. If we could accept these broad ethical norms then any particular culture or religion should be able to find a common ground for common action, while they reinterpret the norms within their own traditions and beliefs.

If we can agree to this, we will respect our environment as individual persons recognizing that our environment includes other humans, the environment in which we live and work, the biological organisms with which we share the planet and the physical environment, and if we base our actions on the factual knowledge that scientific and intellectual study has produced, and act within a common environmental ethical system, then we have a basis for environmental education, discussion and debate about common goals and plans, and common action to resolve global environmental problems. Of course, there is no utopian solution to environmental dilemmas. Such solutions belong only to dreams (Callenbach, 1979). Rather, the task is to create environmentally and ethically sound mechanisms to resolve problems. With such mechanisms we can act in both an environmentally and socially responsible way in the global and local systems in which we live.

ACKNOWLEDGEMENTS

I am grateful for the critical comments of R.J. Berry and Frederick Ferré. Professor Ferré's council to end this paper with a bolder statement of the need for a common respect for the use of factual knowledge and within a generally accepted environmental ethic was especially helpful.

REFERENCES

Callenbach, E. (1979) *Ecotopia*, Bantam Books, New York.
Chase, A. (1986) *Playing God in Yellowstone: the Destruction of America's First National Park*, Atlantic Monthly Press, Boston.
Cronan, W. (1983) *Changes in the Land; Indians, Colonists and the Ecology of New England*, Hill and Wang, New York.
Crosby, A.W. (1986) *Ecological Imperialism, The Biological Expansion of Europe 900–1900*, Cambridge University Press, Cambridge.
Evans, E. (1969) *Essays in Scotch-Irish History*, Routledge, Kegan Paul, London.
Fox, S. (1981) *John Muir and his Legacy, the American Conservation Movement*, Little, Brown, Boston.
Golley, F.B. (1987) Deep ecology from the perspective of ecological science. Environ. Ethics, **9**, 45–55.
Leopold, A. (1949) *A Sand County Almanac* and *Sketches Here and There*, Oxford University Press, Oxford.
Lillard, R.G. (1947) *The Great Forest*, Alfred A. Knopf, New York, pp. 65; 106; 210.
McWhiney, G. (1988) *Cracker Culture, Celtic Ways in the Old South*, University of Alabama Press, Tuscaloosa.
Nash, R. (1982) *Wilderness and the American Mind*, 3rd edn, Yale University Press, New Haven.
Naess, A. (1973) The shallow and the deep, long range ecology movement: a summary. *Inquiry* **16**, 95–100.
Pinkett, H.T. (1970) *Gifford Pinchot, Private and Public Forester*, University of Illinois Press, Urbana.
Quinn, D.B. (1966) *The Elizabethans and the Irish*, Cornell University Press, Ithaca.
Scarce, R. (1990) *Eco-Warriors, Understanding the Radical Environmental Movement*, The Noble Press, Chicago.
Turner, F.J. (1893) The significance of the frontier in American history. Printed in *The Frontier in American History*, 1920, New York, pp. 1–38.
Van Doren, M. (ed.) (1928) *Travels of William Bartram*, Dover, New York.
Vipperman, C.J. (1978) 'Forcibly We Must', the Georgia case for Cherokee removal, 1802–32. *J. Cherokee Studies*, Spring, 103–9.

Vipperman, C.J. (1982) The 'particular Mission' of Wilson Lumpkin. *Georgia Hist. Q.*, **66**, 295–316.

Vipperman, C.J. (1989) The bungled treaty of New Echota: the failure of Cherokee removal, 1836–38. *Georgia Hist. Q.*, **73**, 540–58.

Williams, R.A., Jr. (1990) *The American Indian in Western Legal Thought: The Discourses of Conquest*, Oxford University Press, Oxford.

Environmental attitudes in Germany: the transfer of scientific information into political action* 3

Wolfgang Haber

Wolfgang Haber is Professor of Landscape Ecology at the Technical University of Munich. He is President of the International Association for Ecology (INTECOL) and was previously President of the Gesellschaft für Oekologie. He served as Chairman of the (German) Federal Council of Environmental Advisors from 1985 to 1990.

3.1 INTRODUCTION

A striking scientific experience of the early 1990s has been the contrast between the wealth of ecological knowledge and insights accumulated and promulgated during the previous 30 years and the lasting ecological degradation of the environment, progressing at a great (and even accelerating) pace, and now reaching global dimensions. Even if they did not always have the last proof of the exact pertinent data, ecologists have long been aware of this ongoing deterioration. It is their knowledge of the responses of organisms and ecosystems to impacts and perturbations that

* Adapted from a lecture presented at the 5th European Ecological Symposium in Siena, Italy, September 1989.

Environmental Dilemmas: Ethics and decisions
Edited by R.J. Berry
Published in 1992 by Chapman & Hall, London. ISBN 0 412 39800 1

enables them to warn against hazards and to predict possibly disastrous developments.

Ecologists therefore have an important message to convey from ecology to politics; it is their duty to inject their well-founded ecological awareness into the decision-making still deeply rooted in purely socio-economic ways of thinking and acting, and which as yet pays only marginal or superficial attention to ecological principles.

Human welfare as a general goal of politics can no longer depend on socioeconomic principles alone, but must include, and even be based on, ecological principles of rational use of resources, now often termed sustainable use. But how does this message get across the boundaries between science and politics? As an ecologist who became involved in scientific advisory work to the German Federal Government, I will try to relate some of my personal experiences and insights from this interface between science and politics.

In 1981, the Federal Government appointed me a member of the 'Federal Council of Environmental Advisors', a body established in 1972 at the same time as the West German environmental administration. It consists of 12 members, mostly university professors from different disciplines. There are two biologists or ecologists, two experts from medicine and health sciences, economists, lawyers, engineers, etc. all whom are expected to be 'environmentally-minded'. Each council member is appointed for 3 years and can be reappointed twice, and he or she is paid a substantial salary. The Council has a permanent staff of about 25 people including 15 scientists, led by an executive secretary. The Council meets for one and a half days every month.

I confess that I had mixed feelings when asked to join the Council. I was aware that much additional work would be involved, and I wanted to continue my activities as a scientist and university teacher. On the other hand, I had I felt a serious responsibility to serve the general public and to disseminate my knowledge beyond my personal ivory tower, although I feared my scientific approach would not be compatible with politics. However, I had a somewhat naive conviction that I would be able to 'move' something. So I joined the Council; in 1985 and again in 1987 I was elected chairman; I left the Council when my term expired in 1990.

What can such an advisory body achieve in environmental matters? The Council can choose its subjects, but the Government has the right to ask the Council to investigate subjects selected by itself or by Parliament. The Council has to issue a report on the state of the environment at least once a decade. From the very beginning, the Council decided to treat environmental matters in depth, trying to integrate the viewpoints of different disciplines, and to avoid minority votes. Up to now, this goal has been achieved, which in my opinion is an important general result regarding the difficulty of environmental dilemmas. The Council will not comment on the day-to-day environmental debate or hysteria – although it is often urged to do so – and tries to keep as clear as possible of the environmental ideologies promulgated by anti-nuclear, anti-pesticides, pro-organic farming, pro-solar energy and similar groups. Their arguments, however, are examined as impartially as possible, provided they are amenable to rational analysis.

The main result of the Council's activities has been a series of comprehensive environmental reports (Table 3.1). These are mostly extensive publications, and are regarded as an important collection of environmental knowledge specially selected and 'digested' for both politicians and alert citizens. The reports also contain recommendations for environmental legislation, and for general rules of behaviour towards the environment. The immediate public attention and response to the Council's reports or statements is usually disappointingly small. When a report is presented to the Government, there is a press conference and a television interview producing some more or less accurate articles or comments, plus some largely irrelevant newspaper headlines. The sheer volume of the reports is a deterrent. However, on a longer time-scale, there is more attention, and even some success. New legislation or changes of laws or decrees, and legal decisions, are based on the reports; and frequently members of Parliament cite from the reports in parliamentary debates on environmental matters. It would be tempting to discuss some of the Council's reports in some detail, or to enlarge on some of the

Table 3.1 Reports and surveys published by the (West) German Federal Council of Environmental Advisors

1. Automobiles and Environment, 1973
2. The Wastewater Tax, 1974
3. State of the Environment I, 1974
4. Environmental Problems of the Rhine River, 1976
5. State of the Environment II, 1978
6. Environmentally Hazardous Chemicals, 1979
7. Environmental Problems of the North Sea, 1980
8. Energy and Environment, 1981
9. Forest Decline and Air Pollutants, 1983
10. Environmental Problems of Modern Agriculture, 1985
11. Indoor Air Pollutants, 1987
12. State of the Environment III, 1987
13. Contaminated Sites, 1990
14. Solid Waste Management, 1991
15. General Environmental Monitoring, 1991

* Average of 278 pages.
† The Council publishes a companion series *Background Materials of Environmental Research*. Twenty volumes were available by 1991.

Table 3.2 Major obstacles hampering the transfer of ecological research results into political action

1. Doomsday or chaos prophecy
2. The simplicity–complexity dilemma
3. Ecological fundamentalism
4. The value problem
5. The scale problem
6. Legislative maze
7. Dissension among ecologists

deliberations of the Council members. However, in a book on environmental dilemmas it is more useful and interesting to reflect on the major obstacles which prevent, obstruct or slow down the incorporation of truly ecological arguments or viewpoints, or at least their appreciation, in decisions both of the Council and of decision-makers. I have identified seven major obstacles, which are listed in Table 3.2

3.2 DOOMSDAY OR CHAOS PROPHECY

Many ecologists have convincingly demonstrated that the present use of natural resources, with all the carelessness and harmful side- and after-effects, will inevitably result in disaster. This message, however, is meant as a general warning against resource depletion and environmental pollution, and as a strong call to change the general attitude towards the environment. The arguments are based on extrapolations of current trends of usage and pollution; the aim of the arguments is to challenge these trends, and prevent them from coming true. It is to call for awareness.

However, there are people – among them some ecologists, but mostly environmentalists and adherents of 'green' parties – who seem to revel in apocalyptic visions, either as a debating tactic or from fundamentalist principles (see 3.4). As a result, the average citizen or politician sooner or later becomes inured to the arguments, or judges them wrong or exaggerated. In either case, ecological scientists to whom both sides defer, are blamed as doomsday prophets, and are not taken seriously.

The mass media glory in the doomsday business, and, well aware of the danger producing indifference, frequently change the topics of concern, resulting in 'pollutant of the month' campaigns. Politicians, susceptible and to some extent dependent on the mass media, respond, or pretend to respond, to such campaigns; they tend to ignore scientists who criticize such media spectacles.

Scientists have unwittingly contributed to doomsday prophecy by their own debate about chaos theory, which in itself has little to do with actual environmental problems. Chaos theory is a new approach to systems complexity and order, but has been thoroughly misunderstood by the mass media – which has used it to blame science for misguiding and fooling society. For them, chaos theory is simply a proof of doomsday prophecy.

In this doomsday turmoil, scientists may feel helpless. Our only weapons are rationality, honesty and sobriety; but they are not as effective as we would wish. Scientific illiteracy is all too common, and even if people take a great interest in ecology, they are disinclined to accept its lessons. They are indeed rather unflattering to most people. For example, they may be unhappy to learn from ecologists that humans share with all other consumers (in the ecological meaning) an unappeasable exploitative behaviour. One of the most basic messages ecologists have to convey is as unpleasant as it is true: human life depends on million-fold daily acts of killing – a 'licence to kill', to quote the James Bond movies – and this is an innate and fundamental characteristic of a powerful and highly competitive heterotrophic consumer. All successful consumers have been selected during evolution for their abilities to exploit resources, and this of course holds for humans, too. It is a basic property of our species, but in 'human ecology' it tends to be tabooed; or is it too trivial to be spoken about?

As a matter of fact, evolution has 'produced' effective regulations for resource exploitation. It is one of the lessons to be derived from ecosystem research that the system – and not the individuals or the populations – sets the limits to over-exploitation, thus achieving a steady state. The system has protective properties which are not rooted in the species. But humans have for centuries put all their ingenuity into removing systems' limits to resource exploitation so that they can increase and intensify systems' outputs. The time has now come to accept the lessons of ecosystem research, i.e. to recognize our exploitative nature, and to invest our intellectual and technical skills into supporting the protective and regulatory processes of ecosystems which alone will grant us sustainability and durability. Only in this way can the doomsday syndrome be overcome.

3.3 THE SIMPLICITY–COMPLEXITY DILEMMA

Ecologists are repeatedly embarrassed by the problems of explaining the huge complexity of ecosystems, and of ecological phenomena in general. We even have difficulty in understanding it ourselves, as witness our disputes in professional meetings or writings. It is not surprising that politicians and the general public fail to understand the rules of ecosystems.

Humans prefer linear thinking to network thinking. Of course one can learn network thinking, and most ecologists are well versed in this. However, neurologists tell us there are insurmountable limits. According to Schurz (1989), the adult human brain can dispose of about 10^6 'thought objects'; the average vocabulary is about 10^5, of which 10^4 are in current usage. A fairly reliable estimate of numbers of objects will not exceed 10^4, and the same limit holds for representation of complex systems in the brain. Now if $n!$ is the interconnected arrangement of n elements forming a system, then $8! = 4032 \times 10^4$. This means that the average human brain, if not specially trained, will not grasp systems composed of more than eight elements, and the number of interconnections or interrelationships the brain actually handles will not exceed 10^2, i.e. 100. Many of the ecosystem models or schemes published in textbooks (never mind reports to laypersons) contain more than eight elements!

Furthermore, humans not only use linear thinking, but also prefer a simplistic interpretation of phenomena, seeking a single cause for every event. This explains the great difficulties ecologists have in applying probabilistic interpretations of possible environmental hazards. Even among scientists who ought to know better, a 'mechanistic causality approach' tends to prevail, simply because it fits more comfortably into everyday thinking. This leads to a selective over-valuation of data and results which fit into this causality model, and a consequent underestimate of findings which seem to blur it.

Thus people, including politicians, have split images about environmental reality. They have a background picture of a currently perturbed simple harmony, overlain with an awareness of impending disasters. Complicating rational judgement, most people believe that disasters will hit others first. This is perhaps the only hint of a perception of complexity, and may be a basis for educating people in true ecological thinking.

Politicians demand simple explanations in simple language not exceeding one page. Can ecologists afford to simplify ecological reality? In my work as an environmental advisor, I always tried to use simple language and to simplify complex issues, often struggling to make politicians understand important inter-relationships. I had some success, but more often I was discouraged. Unwillingly I am being faced to conclude that the environment is being spoiled by oversimplification of its complexity as much as by pollution and degradation. The current biodiversity debate is an example of such a conclusion.

The seriousness of the problem can hardly be overestimated. Environmental impact assessments, which have been mandatory in European Community countries since 1985, require (among other things) public participation. This is certainly a good thing, but my experience with public hearings is disappointing. With few exceptions, elaborate, scientifically sound, probabilistic risk or hazard assessments have not been understood, let alone accepted, weighed and decided upon by the public. More often, the hearings are dominated by nimbyism, inability to grasp the real issues, refusal of information, accusations of real or supposed culprits, and cleverness of arguments instead of objective reasoning or adequate weighing of alternatives. However, such failures also come from the inability of scientists to discuss even rather simple problems with laypersons.

3.4 ECOLOGICAL FUNDAMENTALISM

Ecological fundamentalism or dogmatism is not common among ecological scientists, because it is incompatible with scientific reasoning. But it is rife among environmentalists and 'green' movements. However, it is often ecologists who are held responsible for exaggerated or distorted views on the environmental situation which have emanated from fundamentalists, and blamed for one-sided or even wrong political actions derived from them. Such blame is not wholly unjustified, for environmentalists and green parties are the offspring of ecology – even if ecological scientists regard them as illegitimate or unwanted offspring.

What is ecological fundamentalism? Its main tenet is to regard ecosystem or community organization as normative, harmonious, stable and well-functioning, and as a model to organize human society in order to put it back into equilibrium with nature. There are a number of common 'ecological ideologies': small is beautiful, deification of high species diversity, good rural life (with organic farming), untouched nature remaining unchanged – all this combined with a continuous search for evil forces causing ecological perturbations and disasters.

There is one important point to be made about environmentalists and adherents of green parties: they are much more numerous than ecologists, and for politicians in democratic countries they represent a considerable voting power (in some of the German federal states this has exceeded 10%). Moreover, the environmentalists are indirectly – in Germany at least, more strongly supported by the mass media than other political groupings – resulting in the propagation of quite a few environmental reports of a 'fundamentalist' character.

Ecological scientists follow these activities with mixed feelings. On the one hand, environmentalist groups and green parties in Germany have undoubtedly had a positive influence on environmental awareness and activities in the traditional political parties, and on German environmental politics in general. On the other hand, scientists have to react to, and sometimes correct, over-crude simplifications or distortions of ecological reality. To give an example, environmentalists in Germany managed to insert into the draft of a National Park management plan a requirement that species diversity in the park should be actively promoted. But the area in question was a granitic mountain with forests on poor acid soils, which had never been rich in species. Ecologists were able to amend this edict but only at the eleventh hour.

Other examples of ecological fundamentalism in Germany include the proposal to promote organic farming by law, and to ban all waste incineration because waste can be either avoided or recycled. Such proposals are difficult to handle because they are popular, and even sound in principle, but not in implementation. Thus the Council of Environmental Advisors, after a thorough investigation of the waste problem, emphasized the necessity of incineration for the forseeable future, but recommended it with strict conditions and in conjunction with a solid waste tax. The Council was strongly criticized by the environmentalists who accused it of betraying ecological principles.

The political parties in Germany, willing to pursue a good environmental policy, and competing – at least during election campaigns – for the best approaches, are rather susceptible to ecological fundamentalism, and ecologists often have a difficult time ensuring that ecologically sound solutions are carried through. The dilemma is obvious; responsible scientists want to collaborate with the overall aims of the environmentalists, while tempering the latters' enthusiasms. The problem is guarding scientific reality.

3.5 THE VALUE PROBLEM

Most environmental problems have an ecological base but their solution is often not an ecological or even a scientific method; it is a predominantly psychosocial question. Nearly always there are conflicts over priorities. Good ecological research may result in sound information and proposals suitable for individual choices or actions, but many environmental problems concern communities or even society at large rather than the individuals. Discussions and decisions then ultimately come down to value questions, and 'who wants what'.

Social scientists are much more familiar with value questions than ecologists. As Popper (1984) said, you cannot take away value judgements from a social scientist without taking away his personality. But I believe you can do that with a natural scientist. Of course, ecologists are able to give value judgements and will certainly do so; but are they aware of the difference between ecologically and socially acceptable judgements? And are they aware of the bias of the judgement derived from the pecking order of scientific disciplines, with physics and technology at the top and social sciences at the bottom?

The value problem is closely related to ecological fundamentalism (see section 3.4), because values are often formulated and defended with fundamentalist ardour. In such conflicts, the role of ecologists is to provide the disputants with rigorously analysed objective and detailed information. An example of this is the debate in Germany about baby nappies or diapers. In recent years, many parents have switched over to disposable cellulose nappies for babies which are cleaner and easier to use than traditional ones. Of course this resulted in an increase in the already acute solid waste disposal problem. Environmentalist groups started a campaign to persuade mothers to return to old-fashioned cotton nappies that can be reused 20–30 times. They had little success. Then they tried to make the issue a moral one, accusing comfort-seeking mothers of contributing to unnecessary environmental pollution, calling for a special tax on them. A comprehensive examination of the problem revealed this was a one-sided measure, as laundering and sterilizing nappies 20–30 times produces almost the same environmental pollution as the use of disposable nappies provided that waste disposal (including incineration) was done properly. Typically, some environmentalists flatly refused to accept this analysis. The whole debate became a discussion on how much comfort should be conceded to young parents; an environmentally positive result was improvement of disposal of one-way nappies after use, including biodegradability.

Another, more general value problem refers to the ultimate aim of preservation of nature. In Germany there is a plan for the Federal Nature Conservation Act to be amended. The current version of the law states that nature is to be protected as a basic foundation of human life and welfare. In the amendment, environmentalists and naturalists call for protection of nature for its own sake, and seek scientific support in this regard. The Council of Environmental Advisors had to discuss this problem and came to the conclusion that both reasons are human-made because nature, even if protected for its own sake can only 'raise its voice' through humans who will take its part; and this happens already under the current legislation.

3.6 THE SCALE PROBLEM

Ecological scientists can move freely across many temporal and spatial scales – from the Pleistocene or earlier to the present and into the future; or from microsite to the biosphere. But people will not, and even cannot, follow ecologists in such moves, particularly as many ecologists often omit the proper scale of their arguments.

The time-scale of most people is the life-span of adults (about 40–50 years) which can be covered by more or less dependable reminiscences. Much shorter, but much more relevant for environmental problem-solving, is the 'political time-scale' bounded by elections every 4 or 5 years. A 'far-sighted politician' is one able to consider two election periods.

Likewise, the spatial scale of average people is characterized by the area they can grasp by looking and listening, i.e. a maximum of about 20 miles (or 30 km). In reality, most people will determine their spatial dimension by the area of their land property or real estate, and by the area of their town or village.

All smaller or larger scales in time or space are beyond normal imagination, and require special efforts – which are often greatly underestimated – to draw public attention to relevant ecological problems. This explains why both gene manipulation and global warming attract much interest and fear, but do not exert much influence on citizens' everyday behaviour. The scale problem is also one of the main reasons for the failures or slow progress of international environmental politics. As yet, successful environmental politics are based on national (or even smaller) spatial scales, and even international environmental guidelines or agreements, e.g. within the European Community, are in most cases specifically adapted to national scales.

To comply with the scale problem, ecologists have to make very clear the scale of every investigation or action in which they are involved. They also need to emphasize that with expanding scales scientific statements become less reliable and precise because of the complexity problem (see section 3.3). When it comes to transferring scientific results into political measures in favour of the environment, it is important to seek out politicians or decision-makers according to their scale-mindedness. Successful and trustworthy environmental politicians distinguish themselves by an elevated scale-mindedness.

3.7 LEGISLATIVE MAZE

In the Federal Republic of Germany, the European Conservation Year 1970 vigorously promoted general environmental awareness, and the government responded by proclaiming a comprehensive environmental programme. Ministries and other environmental institutions were established everywhere, among them the Council of Environmental Advisors (p. 34). Subsequently, more than a dozen environmental laws were passed by the Federal Parliament, and many specific regulations for environmental improvement have been decreed. Both the Council of Environmental Advisors and environmental activist groups (including the Green Party) pushed the 'classic' political parties into a competition for a better environment.

An ecologist should be highly satisfied with such progress in environmental legislation. Basically he/she is, but there remains substantial criticism. First, implementation is still insufficient, partly because the creation or extension of environmental agencies did not keep pace with the speed of new legislation, and partly because the country's environmental situation cannot be monitored closely enough to detect all changes or impacts. Second, and this is more important from the ecological viewpoint, the many laws and regulations have not been sufficiently coordinated and harmonized to form a coherent system of environmental legislation. The result has been that problems in one environmental sector, such as water pollution, are 'solved' by transferring them into another sector, e.g. soil being polluted or spoiled by sludge disposal.

A particularly bad example of this legislative maze is nature and landscape conservation. The Federal Nature Conservation Act of 1976 decreed 'landscape planning' as an instrument to implement all measures necessary for maintenance and development of natural elements or areas throughout the country. In the same law, another article regulated the handling of environmental impacts on the land and on water-bodies with the aim to avoid, or to mitigate, disturbance or destruction of natural features or areas. The scientific approaches and technical measures required for both landscape planning and impact handling are largely identical, but different agencies or institutions were entrusted with implementation tasks. Some years later, the General State Planning Act ('Raumordnungsgesetz') introduced a fixed general planning procedure ('Raumordnungsverfahren') aiming at an assessment of environmental, economic and social impacts of large public or private projects, again requiring a similar ecological approach for its implementation. Finally, a European Community directive of 1985 obliged Germany to introduce special legislation for environmental impact assessment procedures. The maze was complete, but it was defended by environmentalist groups because of the advantage of trading four different legal instruments against each other to prevent or slow down unwanted developments. They protested loudly when the German government initiated an effective simplification of this complicated legislation for the five new German states (the former German Democratic Republic) in order to promote a quicker unification of the two different administrative systems.

The legislative maze clearly is hampering a sound transfer of ecological knowledge into environmental politics; is wasting expertise and personal strength; and does not help ecologists, who are even blamed for having contributed to this confusion. This may even be true – because of a lack of sound holistic approaches to environmental laws, leading on to the next, and last, major obstacle.

3.8 DISSENSION AMONG ECOLOGISTS

There is hardly a scientific discipline in which all members will agree on basic concepts or issues. Science is driven by doubt, not by certainty. This holds for ecology; but here the degree of dissension appears to be particularly high. It may be a result of the relative immaturity of a rather young discipline still grappling with the challenge to find a common theoretical foundation, and at the same time facilitating its application. However, dissension among ecologists worry just those politicians who are inclined, or even determined, to establish sound environmental politics based on ecological insights and predictions.

Ecologists disputing, sometimes acrimoniously, about a more genuine or a 'more exact ecology' (Grubb and Whittaker, 1989) should always consider the fact that their science has rather suddenly been drawn into the focus of public attention concerned about the credibility of ecological concepts or applications. And ecologists must be wary of press or television correspondents taking note of their disputes, and even taking sides in them; science is no longer an internal affair of scientists.

Among current arguments, the holism-reductionism debate is still one of the most important. It should be quite clear that environmental planning or management requires a holistic approach based on ecosystem ecology, and this is what decision-makers expect, even if they have never heard about holism. But the instruments ecologists apply to exercise that holistic approach are reductionist ones: models, plans, maps, charts, all of them reducing the complex ecological reality to some tractable and intelligible features. Of course these have to fit, and be tested against, reality, if possible using an iterative procedure. Thus holism and reductionism can be regarded as two sides of the same ecological coin. But this coin has to be looked at, or examined, from both sides or, in modern scientific slang, a top-down and a bottom-up investigation of the system has to be exercised by turn. This is not at all new; 100 years ago ecologists had an implicit concept of the ecosystem as an entity persisting through time (O'Neill and Giddings, 1979).

Another controversy comes from the gap between field and systems ecologists, or between experimental and modelling approaches. Both complement the other in the investigation of complex structures and processes, but it is often necessary to educate them. Field work producing large quantities of data may be of just as little use as computer work generating lots of questionable, but attractive models. What is needed is the minimum amount of data that allows one to understand the system under study or the problem being formulated. This means that data collection should be structured from the beginning by an adequate conceptual scheme or problem formulation. This requires cooperation, and not controversies, between field ecologists and modellers. To solve the spruce budworm problem in eastern Canada, a million dollars were spent on baseline studies to describe the ecosystem dynamics involved. But it was the qualitative knowledge of a few experienced scientists who knew how budworms, trees and birds interact, and who had an intuitive knowledge of long-term system behaviour, that provided the keys for the solution of the problem (Miller, 1985).

The transfer of ecological research into political action for improving the environmental situation cannot be done without planning and modelling. But these activities are viewed with suspicion by a number of ecologists who feel that they are incompatible with 'good science'. Ecologists who engage in planning and modelling are often harshly, and publicly, criticized. There are cases where such criticism is justified, because of apparent lack of scientific rigour. My own reservations about planning and modelling spring from a more fundamental reflection: environmental planning pretends to know the future development and behaviour of complex systems, but the predictive power of ecologists is at best restricted; moreover, ecological planning contradicts the theory of self-organization and self-maintenance of systems. It is this underlying problem that urgently needs thorough exploration involving also economics and social sciences (Kanitscheider, 1991).

The urgent application of ecological knowledge to the solution of environmental problems should convince ecologists of the necessity for stronger efforts in building a unified science. The First International Congress of Ecology, held in The Hague in 1974, had the general topic 'Unifying concepts in ecology'; however, the time did not then seem to be ripe for ecologists to speak with one understandable voice. At least we could agree on two different, but equally necessary, research strategies: basic research seeking knowledge and applied research seeking solutions to problems. We are never sure if research will unveil scientific truth, but the results can be of heuristic value. A careful choice of observation sets, proper consideration of levels of organization and pertinent temporal and spatial scales (O'Neill *et al.*, 1986), and a planned research structure in which all expected data will fit, are necessary prerequisites for overcoming ecological controversies.

3.9 CONCLUSIONS

The seven major obstacles which I have found are hampering the transfer of scientific ecological information into environmental policy may be reduced with time and insight but will never disappear. Ecologists do well to keep them in mind. Moreover, they should not forget that in influencing decision-makers and politicians, they are competing with economists, lawyers, social scientists, and the mass media. It is still not uncommon to leave this competition as losers; but they may take comfort in ending as winners at the final run-in. We should perhaps accept this somewhat pragmatic attitude as the outcome of our seriousness and thoroughness. As McIntosh (1986) wrote: 'Ecologists have for decades been offering advice at least as soundly grounded in experience and theory as that offered by economists . . . They have been much less successful . . . in getting a hearing for their advice . . .'

And there is the yearning for social equity. Rawls (1971) in his *Theory of Justice* aptly wrote: 'If the world were fair, we would willingly enter it – but depending on location at a given point in time or time at a given location.'

What a fitting ecological statement from the pen of a lawyer! 'Location' (= site, ecotope) involves a given set of natural resources, both renewable and non-renewable – but ecologists know that the spatial distribution of these sites and sets across the continents is uneven, hence 'unfair' to humans seeking equal chances or equal rights. But this uneven resource distribution is most probably the main reason of the origin of biodiversity – doesn't it hold for humans too?

REFERENCES

Grubb, P. and Whittaker, J. (1989) *Towards an Exact Ecology*, Blackwell Scientific Publications, Oxford.

Kanitscheider, B. (1991) Selbstorganisation in komplexen Systemen. *Universitas*, **46**, 751–60.

McIntosh, R.P. (1986) *The Background of Ecology*, Cambridge University Press, Cambridge, p. 332.

Miller, A. (1985) Technological thinking: its impact on environmental management. *Environ. Management (New York)*, **9**, 179–90.

O'Neill, R.V. and Giddings, J.M. (1979) Population interactions and ecosystem function, phytoplankton competition, and community production. In *Systems Analysis of Ecosystems* (eds G.S. Innis and R.V. O'Neill), International Cooperative Publishing House, Fairland, Maryland, pp. 103–23.

O'Neill, R.V., DeAngelis, D.L., Waide, J.B. *et al.* (1986) *A Hierarchical Concept of Ecosystems*, Princeton University Press, Princeton, New Jersey.

Popper, K.R. (1984) *Objektive Erkenntnis – Ein evolutionärer Entwurf*, Hoffman und Campe, Hamburg.

Rawls, F. (1971), *A Theory of Justice*, Oxford University Press, Oxford.

Schurz, J. (1989) Das Gehirn als System. *Naturwiss. Rundschau*, **42**, 345–53.

Case study: air quality 4

T.M. Roberts and J. Sheail

Mike Roberts is the Director of the Natural Environmental Research Council's Institute of Terrestrial Ecology (South). Previously he was Section Head with the Research Division of the Central Electricity Generating Board, responsible for research into 'acid rain'. He has edited a number of books, including Ecological Prospects of Radionuclide Release *(Blackwell Scientific, 1983) and* Planning and Ecology *(Chapman & Hall, 1984).*

John Sheail also works with the Institute of Terrestrial Ecology (South). He is a distinguished biological historian. He has written several books, including Nature in Trust: The History of Nature Conservation in Britain *(Blackie, 1976),* Seventy-five years in Ecology: The British Ecological Society *(Blackwell Scientific, 1987) and* Power in Trust: The Environmental History of the Central Electricity Generating Board *(Clarendon, 1991).*

4.1 INTRODUCTION

The UN Conference on the Human Environment, held in Stockholm in 1972, may be taken as the beginning of the current high profile of the environmental sciences in international politics. Never before had demands come from so many different quarters for *scientific* evidence to fuel the political arguments for improving the quality of the environment. At the top of the agenda for the Swedish hosts was publicity for a recently-completed report on the way in which rainfall acidity, caused by sulphur dioxide from the UK and other industrialized countries, was bringing about the destruction of Scandinavian lakes and forests (Royal Ministry, 1971).

Environmental Dilemmas Ethics and decisions
Edited by R.J. Berry
Published in 1992 by Chapman & Hall, London. ISBN 0 412 39800 1

In the intense debate that followed over the subsequent two decades as to how sustainable development can be achieved on a global scale, policy-makers and the media have increasingly sought more information, particularly on the pathways and effects of air pollutants. This concern for the planet's atmosphere caught the scientific community and regulatory agencies ill-prepared. Neither had thought – or were organized – to work on so extensive a geographical scale. The basic premise of emission regulation, namely that 'the answer to pollution is dilution' was challenged. It was no longer simply a case of deciding how much higher a chimney might be built, or strengthening the existing powers of a government department. A completely new international dimension of environmental research and management was shown to be required. The recently-formed Commission of European Communities (CEC), along with other international organizations, such as the WMO, WHO, UNECE, UNEP and OECD, became the driving force in determining the scope and character of further air-pollution research and legislation.

The dilemmas for both politicians and scientists, arising from the 1972 Stockholm Conference, were particularly acute in the UK. It seemed scarcely credible that pollution arising from local sources could be creating problems on a regional scale. The Clean Air Act of 1956 had significantly reduced air pollution in cities in the UK. Even if a long-range transport of emissions could be demonstrated, was it sufficient to justify the costs of emission controls? International politics seemed to be outrunning scientific understanding, falsely linking emissions in one country with impacts in another. Could the natural environment really be so sensitive that 'achievable' emission reductions, as determined by politicians and based on scientific principles, might not be enough to ensure acceptable environmental quality?

This chapter begins by looking more closely at the acid rain debate, and the expertise and experience which UK scientists and policy-makers brought to the debate as it has developed over the past two decades. Drawing on this case study, the chapter concludes by outlining more generally how a global agenda for pollution control has developed and the difficulties created by reduced time-scales for resolution of scientific uncertainty and the increasing political importance being attached to the 'precautionary principle' in accommodating public suspicion and hostility, as to what *might* be happening to the environment.

4.2 THE ACID RAIN DEBATE

Progression from hypothesis to controversy to consensus – with many disagreements between scientists, misunderstandings and deliberate

Table 4.1 Acid Rain Development in the UK – A Select Chronology of Events

Year	Event
1956	UK Clean Air Act
1963	Nuclear Test Ban Treaty
1970	European Conservation Year. Appointment of UK Royal Commission on Environmental Pollution
1972	UN Conference on the Human Environment
1977	OECD Report on Long Range Transport of Air Pollutants
1979	UNECE Convention on Long Range Transboundary Air Pollution ratified
1982	Second Stockholm Conference
1983	West German programme to reduce sulphur dioxide emissions by half in 10 years
	Formation of the 30% Club (not including the UK)
	Surface Water Acidification Programme launched by CEGB/British Coal
1984	House of Commons' Environment Committee Report on Acid Rain
1986	Chernobyl nuclear power station accident
	Announcement that flue gas desulphurization to be fitted to all new coal-fired stations and to 6000 MW of existing high-merit plant
1987	Announcement of low NO_x burners to be fitted to 12 largest power stations
1988	EC Directive on Large Combustion Plant
1989	Montreal Protocol on CFC emissions
1991	UK White Paper, 'This Common Inheritance'
	UK Government announce plans for an Environmental Agency
1992	UN Conference on Environment and Development

obfuscations by the media and politicians along the way – follow a predictable pattern for any scrutiny of a major environmental issue (Table 4.1). The original assertion by Oden (1968) that sulphur levels in Swedish surface waters had increased, as a result of the long-range transport of sulphur dioxide emissions, generated a climate of scientific and political challenge. Many of these were led by the generally dismissive UK Government, supported by the electricity industry. In order to understand the basis of these challenges, reference must be made to the considerable advances made in environmental research and development in the UK electricity utilities which operated under strict statutory regulation, before the Stockholm Conference. They provide a context for the slow response to demands for a new approach to air pollution problems that were rapidly emerging on the international front.

4.2.1 BEFORE STOCKHOLM

The electricity industry is one of the most closely-regulated industries in the UK. Ever since 1909 it has been illegal to build or extend a power station without statutory consent. So far as air pollution is concerned, the question resolves itself into three issues: the elimination of smoke, of grit and dust, and of sulphur dioxide fumes, with the first and second problems largely resolved by the late 1920s. Experience indicated that hardly any nuisance would arise if mechanical grit arrestors were installed in the flues, and chimneys were built at least two-and-a-half times the height of surrounding buildings (Sheail, 1991).

Sulphur emissions proved less tractable. To secure consent for further phases of the Battersea power station being built in the 1930s, a flue-gas washing process was developed from the laboratory to working-plant scale within 5 years, whereby the sulphur was removed by sprays of water from the river Thames to which chalk was added. An improved process was installed a few years later at the Fulham power station, and further major advances were made in a third station, the Bankside power station, built immediately after the war. It was 'a notable and pioneering achievement' in research and development. It was also a blind alley, in as much as there would never be enough river water to clean the gases and remove the effluent from the size of power stations contemplated for the 1950s and 1960s. Furthermore, washing cooled the boiler gases to such an extent that the plume descended more quickly to the ground, causing concentration of any residual sulphur dioxide in the vicinity of the plant.

An alternative approach was advocated by the industry's expert witnesses, who appeared before a succession of public inquiries convened by the Minister of Power, in the course of considering applications for the construction of further power stations in the 1950s and early 1960s. It was contended that if the chimneys could be built to as great a height as possible, and the (unwashed) gases were discharged at temperatures of, say, 100°C, the plumes would rise quickly to a level where there was adequate dispersion before descending to the ground. As in the early 1930s, once the need for research and development was recognized, results came quickly. The monitoring studies showed that even the most optimistic predictions of the CEGB were realized. The daily average sulphur dioxide concentrations recorded at six 2000 MW power stations indicated that each had a negligible effect on local concentrations (Clarke and Spurr, 1976). The key to this success had been the dispersal of the gases from a single, multiflue, tall chimney. The structure represented less than 1% of the total capital costs of a modern coal-fired power station (CEGB, 1981). It was a further 'notable and pioneering achievement', in the course of which the industry itself became a leading participant and authority in what had come to be known as the environmental sciences.

4.2.2 AFTER STOCKHOLM

With hindsight, it was regrettable that the Swedish report on regional acidification of precipitation, soils and surface waters, and the consequent decline of forests and loss of fisheries (brought to the attention of the international audience at Stockholm in 1972) was presented as 'conclusions' rather than as a series of hypotheses. It encouraged the Scandinavians to adopt, on the one hand, an almost evangelical role in promoting the need for further controls over emissions and, on the other hand, a defensive position in the inevitable scrutiny of their research findings that followed the Conference.

Phase I of the Norwegian SNSF Project on the 'Impact of acid precipitation on forest and freshwater ecosystems' covering the years 1972–75, supported the hypotheses. It provided the framework for ecological investigations that were to last a decade before consensus was achieved. The report on the first phase (Braekke, 1976) described the increasing acidification of lakes and streams in southern Norway, an area of high sulphur and nitrogen deposition and slowly-weathered granite bedrock. Fish mortality was ascribed primarily to acute periods of low pH linked mainly to snowmelt. Short-term experiments on soil columns showed that simulated acid rain accelerated soil cation leaching, but the extent of declines in forest growth appeared minimal. These conclusions were to prove correct in outline, but not in detail or time-scale. The main problem arose from the earlier, spurious balance between direct sulphate input in bulk precipitation and sulphate output from lakes. Such a balance disregarded key processes in the catchment, controlling both inputs (including dry and occult deposition and evapotranspiration) and runoff chemistry (particularly exchange processes in soils).

There soon followed a report from the Organization for Economic Cooperation and Development (OECD) on long-range transport of air pollutants (OECD, 1977). This was particularly important in attempting, for the first time, to quantify 'who was doing what to whom'. The report's principal conclusion was that most of the sulphur deposited in Norway was derived from emissions elsewhere in Europe, and from the UK in particular. This report presented the UK with a *dilemma* – had the tall-stack policy protected local communities at the expense of the natural environment in remote locations? The financial implications of such a finding were substantial for the CEGB, and a research programme was rapidly mounted, encompassing dispersal processes and atmospheric chemistry, pathways through terrestrial ecosystems, and effects on freshwater chemistry and fisheries. The beginnings can also be discerned of a research interest on the part of the UK Government, through the Department of the Environment (DOE), and in one of the country's five research councils, the Natural Environment Research Council.

Unfortunately, much of this research was viewed as a 'challenge' to the Scandinavian 'conclusions'. Scientific debate became dogged by antagonistic views. However, the challenge made consensus imperative and, as a result, fundamental knowledge of many aspects of atmospheric chemistry and ecological processes was significantly advanced.

It was particularly important that further European-wide initiatives should take place. Fortunately, the OECD initiative was followed by another under the aegis of another international body, the UN Economic Commission for Europe (UNECE). The EMEP programme (supported in part by the UK Government) sought to model and evaluate the long-range transport of pollutants. As more data became available, the percentage of sulphur deposition attributed to 'background' deposition in Scandinavia was nearly doubled, with proportional reductions in the amounts derived from each European country (to less than 10% from the UK). The UK research effort also made significant progress. It rebutted the charge that tall stacks 'exported' pollution, thereby preserving the local environment at the expense of more distant parts. It was found that emission of sulphur dioxide from 'tall chimneys' increased the long-range transport by less than 15% compared with emissions from low-level sources. However, the original hypothesis – that long-range transport of sulphur and nitrogen emissions made a significant contribution to deposition in high rainfall areas in southern Scandinavia – remained unshaken.

Several attempts were made to establish a causal relationship by linking temporal trends in the chemistry of precipitation and surface waters with the increase in emissions of sulphur dioxide and nitrogen oxides in the postwar period. While trends for a few sites in Europe and North America were reasonably convincing, interpretation of the records of the European Air Chemistry Network for North Europe, between 1955 and 1975, varied according to the period of observation and the site selected. However, spatial trends showed acid deposition was higher in regions with high emissions. 'Episodes' of acid rainfall in receptor areas were back-tracked by trajectory analysis to the passage of air masses over areas of high emissions.

Attempts were made to emphasize the importance of natural organic acids in surface water acidification but it became clear that sulphate was the major anion linked with acidity. Early observations that fish-kills were caused by acute episodes of low pH gave way to the view that dissolved inorganic aluminium was an important toxic factor in oligotrophic waters below pH 5–5.5. The primary effect of aluminium was thought to be the disruption of sodium regulation at the gills. This effect was moderated by calcium. Thus, by the early 1980s, the combination of low pH, low calcium and high inorganic aluminium in surface waters was seen to be the critical factor.

The hypothesis that acidity in lakes and streams was caused mainly by direct inputs of acid precipitation and snowmelt runoff had been challenged by, among others, a Norwegian geologist (Rosenqvist, 1978). As a consequence, it was established that apart from high runoff episodes, much of the acidity in surface waters was derived from the catchment. Rosenqvist and others went on to argue that natural processes, such as litter decomposition, were a much larger source of hydrogen ions than acid deposition. A key paper by Johnson and Cole (1980), outlining the significance of mobile anions in driving soil leaching, laid the basis for rebuttal of this argument. Organic acids could indeed lead to acidification of upper soil horizons, but were decomposed lower down the profile. Mobile inorganic anions (such as sulphate, nitrate and chloride) were required to drive cations, including hydrogen and aluminium, out of soils into surface waters. Changes in land use, such as afforestation, could alter the fluxes of mobile anions and modify the availability of associated cations, but could not generate acid waters in the absence of mobile anions. Neutral salts, such as sodium chloride in seaspray, could produce acid episodes if the cation displaced hydrogen from the mineral exchange complex, but these effects were transitory.

Attempts to establish historic trends for surface water chemistry and fish losses gradually led to the conclusion that acidification had been taking place over a much longer time-scale than previously thought. This was confirmed by studies of the diatom record in lake sediments, which could be related to historical trends in water acidity. Several studies in south Scotland, Wales and Norway showed that acidification had been taking place at different rates since the beginning of the Industrial Revolution, as a result of the accumulation of acid deposition. The rates reflected such variables as vegetation type, soil buffering capacity, geology and hydrology (Howells, 1990).

4.2.3 THE SECOND STOCKHOLM CONFERENCE

The debate was becoming more political. In July 1982, the Swedish Government organized a meeting to review progress since the historic Stockholm Conference held 10 years earlier. The UK Government responded by ensuring that representation was at Ministerial level. The electricity industry in the UK regarded the Conference as a transparent move to apply further political pressure. In that, it succeeded. Even though the international scientific debate had just reached its most intense and bitter 'high' point, a political 'consensus' was adopted, whereby, 'for any increase in acid-sulphur deposition, there would have to be either an increase in soil acidity or water acidity, or a combination of both'.

In the absence of a quantitative understanding of acid deposition and fishery decline, the Swedish Government proposed that sulphur deposition in the southern counties of Sweden should be reduced to a level not greater than in the north of Scandinavia where fisheries still flourished. Such an adjustment would have required a 75% reduction in emissions from much of Europe, including the UK, causing energy costs to rise by more than 15%. The proposal brought into sharp focus the second *dilemma* – the value of the fisheries was orders of magnitude less than the costs of controlling sulphur emissions. Was it justifiable to protect a relatively small fishery in southern Scandinavia at a cost in excess of £10 billion for Europe as a whole? Was it not more reasonable to introduce comparatively inexpensive 'palliatives', such as liming, at the points of impact? The consequences of the 'polluter pays' principle were now in sharp focus.

While others argued that a value could not be placed on 'environmental quality', it is likely that the matter would have rested there if it had not been for the warnings of some scientists in West Germany in 1980–81, which drew attention to the unprecedented decline of Silver fir and Norway spruce in high-altitude forests. The 'disease' was presented as progressive and irreversible. Acid rain was again invoked as the main causal agent. Based on intensive nutrient-cycling studies at the Solling Forest, Ulrich *et al.* (1980) predicted that acid deposition would lead to soil acidification, aluminium toxicity, damage to tree root structure, nutrient deficiency, and forest decline. By 1982, observations of similar 'declines' were reported elsewhere in Europe, the USA and Canada.

Most of the participating countries at the 1982 Stockholm Conference temporized by affirming their support for the Convention on Long-range Transboundary Air Pollution, drawn up by UNECE in 1979. They undertook specifically 'to limit and, as far as possible, gradually reduce and prevent long-range transboundary air pollution'.

All parties in the UK recognized the political power of scientific consensus. The DOE stepped up its research programme and convened a series of Expert Review Groups to assess the effects of acid deposition in the UK (DOE, 1983). The objectives were to understand the scale and extent of acid deposition and the processes by which emissions brought about atmospheric reactions, deposition and impacts. Long-term monitoring programmes were devised, covering air pollutants, soils, waters, trees, buildings and materials. With much media attention, scientists, funded by the DOE and NERC, demonstrated how the 'Scandinavian' problem of acid deposition and acid lakes was also occurring in the high-rainfall areas of south-west Scotland, the Lake District and upland Wales.

In West Germany, extensive forest-health surveys in 1983 indicated that 35% of the forest area was suffering excessive foliage loss. There were by then almost 200 hypotheses to explain forest decline. Many did not survive even the most superficial scrutiny. The two most credible mechanisms involved air pollution. In one case nutrient deficiency was caused by a combination of soil acidification and aluminium toxicity. In the other, it came about through direct ozone damage and foliar leaching by acid deposition. A West German Government Advisory Group in 1983 could not reach a consensus as to the single most important cause of decline, but agreed that 'air pollution could not be eliminated as a contributory factor' (Anon, 1983). Confronted by the spectre of acute damage to forests throughout central and eastern Europe, the West German Government announced a programme to reduce sulphur emissions by half within 10 years, and the gradual introduction of catalytic convertors to reduce nitrogen oxides and hydrocarbon emissions from vehicles.

The complete reversal in the position of West Germany, the second largest exporter of sulphur dioxide in Europe, left the UK even more exposed. The DOE and Department of Energy became convinced of the need to demonstrate active exploration of ways to achieve reasonable reductions of emissions. The Inspectorate of Industrial Air Pollution notified the CEGB that flue gas desulphurization would be regarded as the 'best practicable means' for controlling sulphur emissions at any *new* plant that might be built.

The CEGB approach was described as one of 'visible movement with minimum penalty'. Research focused on emission-control technology and assessment of the different environmental impacts. Studies on new 'clean technology' concentrated on the development of combined gas turbine/steam turbine cycles, based on either pressurized fluidized bed combustion (PFBC), or coal gasification. Impacts research took a novel approach. In addition to its own expanding in-house programme, the CEGB invited the Royal Society of London, and the Norwegian and Swedish Academies of Science, to organize an independent 5-year programme of research on Surface Water Acidification (SWAP). The CEGB and British Coal established a £5 million Fund to support the research. Critics dismissed the initiative as an attempt to buy time and integrity. For the CEGB, it was evidence of a determination to find solutions that were effective, and not just 'a crippling burden on electricity consumers'.

Committed to a substantial emission-control programme, the next priority for the West German Government was to ensure that other nations made similar sacrifices. At a UNECE conference in June 1983, nine countries (including West Germany, Sweden, Norway and Canada) proposed a 30% reduction in sulphur dioxide exports in the decade 1980–90. They became known as the 30% Club. The UK, France and Italy resisted the proposal. The UK Government argued that the base year of 1980 was arbitrary. While it represented a peak value for many countries, the UK had been progressively reducing sulphur dioxide emissions since 1971, and by 1990 would have reached a reduction of 40% below peak emissions. France claimed that the scale of its nuclear power programme made any further abatement of sulphur dioxide emissions unnecessary.

About the same time, the European Commission put forward two draft Directives. The first was to harmonize procedures for authorizing pollution controls on new plant, based on 'the best available techniques not entailing excessive costs' (BATNEEC). The second Directive set a numerical limit for sulphur dioxide (SO_2) and oxides of nitrogen (NO_x) emissions from new plant and sought substantial decreases from existing plant. The second dilemma was now in sharp focus – the UK argued that international action was now outstripping scientific understanding. More definitive research was still needed to establish the real dimensions of the problems, and the most effective solutions.

The position of industry in the UK was, however, becoming more exposed. In February 1984, France announced a 50% reduction in sulphur emissions. In the same month, the Royal Commission on Environmental Pollution acknowledged that further research was needed on the nature and effects of acid deposition, but insisted that the CEGB should also carry out trials using various forms of sulphur dioxide abatement. In the early summer of 1984, the issues were debated before the Prime Minister at Chequers. For the CEGB, there were two issues, namely whether flue gas desulphurization should be fitted to new plant and, much more importantly, to existing plant. The Department of Energy and CEGB argued that the cost of retrofitting (at £7–8 per tonne of coal burnt) could not be justified by the limited 'proven' extent of effects on fisheries and forests. The lack of a consensus on the forest decline issue seems to have been the key point in the resulting decision not to adopt emission controls.

Meanwhile, the Environment Committee of the House of Commons was taking evidence on the issue. It learnt of how a forest-health inventory had shown even greater and more extensive losses of foliage in West Germany. Pressure groups in the UK gave much publicity to the remarks of a German forester who claimed to see similar signs of forest decline caused by acid rain in the UK.

The report of the Commons' Environment Committee was published in July 1984. It cited the magnitude of lake acidification in Scandinavia and of forest decline in central Europe as a warning of what could happen in the UK (Parliamentary Papers 1983–84). A particularly controversial section linked acid rain to the accelerated rates of corrosion observed on many important buildings in the UK. The CEGB protested that the report's value, in terms of scientific analysis, had been undermined by its adoption of 'acid rain' as a blanket term for all gaseous air pollutants. The case for controls over sulphur emissions was being driven by a concern over the effects of all types of pollution, whether acidic or not (Howells, 1990). In its response to the Committee's report, the Government drew attention to the fact that UK emissions of sulphur dioxide had fallen by 40% since 1970 (from 6.09 to 3.72 million tonnes).

The UK could not, however, shake off the fact that its sulphur dioxide emissions were still the highest in western Europe. Backed by the Nordic nations, it was now easy for pressure groups within the UK to attach the label 'dirty man of Europe' to the UK (Elsworth, 1984; Pearce, 1987). Matters were not helped by the fact that the same pressure groups and the Forestry Commission could not agree on the state of the UK forests. The Commission accepted that crown thinning was widespread, but insisted that this was from natural causes. Unfortunately, there were no data to support that contention; the Forestry Commission was seen to be dragging its feet on research. The scientific community had no option but to sit on the fence – the potential interactions between air pollutants and other stresses had not been sufficiently explored (DOE, 1988). The Nature Conservancy Council broadened the debate by drawing attention to the implications of acid deposition for nature conservation in the UK (Fry and Cooke, 1984). The growing isolation of the CEGB's position was particularly evident following the release of its own 'Acid rain' video in July 1985. From many quarters, there was hostility and ridicule for its rehearsal of the many 'uncertainties' as to the causes and effects of acid rain.

4.2.4 RESOLUTION

By early 1986, the pieces of the scientific jigsaw puzzle were falling into place. The investments made by the CEGB, DOE and the Natural Environment Research Council in research were bearing fruit.

Even more significantly, major revisions were required to forecasts of load demand in the electricity industry. Until the mid-1980s, it had seemed highly unlikely that emissions would increase substantially. The country was in recession, and the nuclear sector was expected to provide any additional capacity required. However, that position changed as forecasts of electricity demand rose, and the development of nuclear power was further delayed. Far from emissions declining in the 1990s, there was, as the Board's own predictions made clear, the likelihood of an upward trend for at least a few years. If that increase were to be averted, limits would have to be imposed on emissions from both new and existing stations.

In March 1986, the CEGB announced that flue gas desulphurization would be fitted to new coal-fired stations, thereby ensuring a steep decline in emissions beyond the year 2000. The decision was also taken internally to embark on a retrofit programme. Some 6 months of intense discussion followed, in which the scientific evidence of environmental impacts played a crucial part in justifying to Government the large-scale expenditure required, both in terms of capital and running costs to the industry and, therefore, consumers of electricity (Chester, 1986).

In 1983, a 5-year international research programme had been launched, under the aegis of the Norwegian Institute of Water Research, known as Project RAIN (Reversing Acidification in Norway). Results obtained in 1985 demonstrated how the sulphur store in soils regulated the response of surface waters to sulphur deposition. If the soil sulphur reservoir was in equilibrium with current deposition, the output of sulphur would approximate the input. In consequence, decreasing the annual deposition would not have a proportionate effect. It would simply make a small reduction in the size of the sulphur reservoir possible. If increased, it would add to the size of the reservoir, and prolong the problem. Modelling studies indicated that the decrease in acidity, plus inorganic aluminium in soil drainage, might take decades to occur as the sulphur store in some soils was so large. The replenishment of calcium by mineral weathering or atmospheric input would have taken even longer. It confirmed the need to add lime to such soils and water. Initiation in 1984 of the DOE liming studies at Llyn Brianne and the Esk/Duddon, and the CEGB Loch Fleet Liming Project in south-west Scotland, was justified.

By 1986, significant progress had been made in explaining forest decline, although all the causal links had not been explained. Rather than *Waldsterben* (forest *death*), there were several types of *decline* taking place in West Germany, each with its own unique combination of causal factors. The yellowing symptoms on spruce and fir in high-altitude forests were the result of magnesium deficiency. Experimental treatments with ozone and acid mist had failed to reproduce the symptoms. Irrefutable evidence of soil acidification in Sweden (Hallbacken and Tamm, 1986) had raised the priority for soil acidification research. Attention was drawn to the potential effects on soils throughout Europe, and the considerable economic implications. Field observations in West Germany confirmed that the nutrient deficiencies were caused by acidification of forest soils. However, experimental work had largely eliminated aluminium toxicity as a significant factor directly in root death (CEGB, 1987). In the UK, the Forestry Commission demonstrated that crown thinning was not a result of magnesium deficiency, and there was no simple spatial relationship between foliar loss and air pollution (Innes *et al.*, 1986).

The long-standing insistence of the CEGB that there was no simple correlation between emissions from power stations and the fate of fisheries and forests had been borne out. Recent increases in acid deposition were significant, but some lakes had been acidifying for nearly 200 years. Acute acid episodes could cause fish mortality, but many fishery declines in acid waters were associated with long-term aluminium toxicity. Surface water chemistry depended on catchment processes. Forest decline was a complex phenomenon with no single cause–effect relationship to acid deposition. The perspectives that had emerged were accordingly very different from those envisaged by the pressure groups and some sectors of the scientific community at the time of the unilateral actions of the West German Government in 1983.

Nevertheless, any further increase in emission levels could not be accepted. Models of surface water acidification predicted that such an increase in surface deposition would both extend and prolong acidification (Chester, 1986). In September 1986, the DOE announced that it had authorized the CEGB to implement its proposals to retrofit 6000 MW of high-merit plant between 1988 and 1997, at a cost of nearly £800 million. In May 1987, the Government endorsed plans by the CEGB to install low-NO_x burners at the 12 largest stations at a cost of £170 million over 10 years, with the aim of reducing emissions by 30% per station.

These moves predated the EC Directive on Large Combustion Plant, which emerged in its final, agreed form in June 1988. Emissions of sulphur dioxide (SO_2) would be reduced from new plant by 80–95%, and by up to 50% for NO_x. For existing plant in the UK, the Directive required phased reduction of sulphur dioxide from 1980 levels by 20, 40 and 60% by 1993, 1998 and 2003 respectively. No_x emissions were to be reduced by 15 and 30% by 1993 and 1998 respectively. International agreements were also reached to install catalytic convertors on motor vehicles to limit NO_x and hydrocarbons – the precursors of photochemical oxidants.

By that time, research on forest decline had achieved a reasonable consensus (Roberts *et al.*, 1989). The common synchronizing factor in Europe had been the droughts of 1976, 1983 and 1984. The extent of foliar loss had stabilized in subsequent years. Predictions of 'catastrophic' forest decline had not been realized. Sulphur and nitrogen deposition were only significant in exacerbating nutrient deficiencies. The effects of magnesium deficiencies in up to 20% of the forest area in West Germany had been increased by tree uptake and leaching by the mobile anions sulphate and nitrate. Photochemical oxidants did not play a significant role in the acceleration of nutrient leaching. The deficiency did not occur in the western seaboard of such countries as the UK and Norway (in spite of high acid deposition) because of replenishment by sea salts in rainfall. The importance of ammonia emissions from agriculture was recognized for the first time. Linked to accelerated dry deposition of sulphur dioxide, they produced, in the Netherlands, nutrient deficiencies through soil acidification.

Not all the issues in the 'Acid Rain' debate have been resolved. There remains the third *dilemma* that the objectives set by the political process, and based on scientific principle, may not be achievable. Soils in some regions may be so poorly buffered that the reduction in emissions required to reverse acidification might be beyond reach. Resolution of the issue has focused on the 'Critical Loads Concept' (Nilsson and Grennfelt, 1988 (Fig. 4.1). Scientific evaluation of the buffering capacity in a receptor area, based on geology, soils and water chemistry, can be used to predict the threshold for acidification (critical load). Political judgement then sets a 'Target Load' with either a margin of safety or a degree of unavoidable but 'acceptable' damage dictated by cost-benefit analysis of controls (Fig. 4.2). Development of new computing techniques for processing large spatial databases (Geographical Information Systems), means that pollutant deposition and regional acidification sensitivity maps can be superimposed to identify areas of 'critical loads exceedance'. Development of GIS techniques for evaluation of the implications of different emission control options has been pioneered in the UK (Bull, 1991). For example, acid deposition over 8% of the UK land area will still exceed the critical load even after the 60% reduction in sulphur dioxide emissions demanded by the EC Directive by 2005. These techniques will be extended to a European scale as the basis of scientific input to the UNECE re-evaluation of emission control objectives in 1994.

There also remains a fourth *dilemma* in that the time-scale for recovery of acidified soil and surface waters probably extends over decades. In terms of accountability, what credibility will be based on scientific judgement if time-scales for recovery extend beyond a generation? It could be argued that the political momentum chose to ignore the

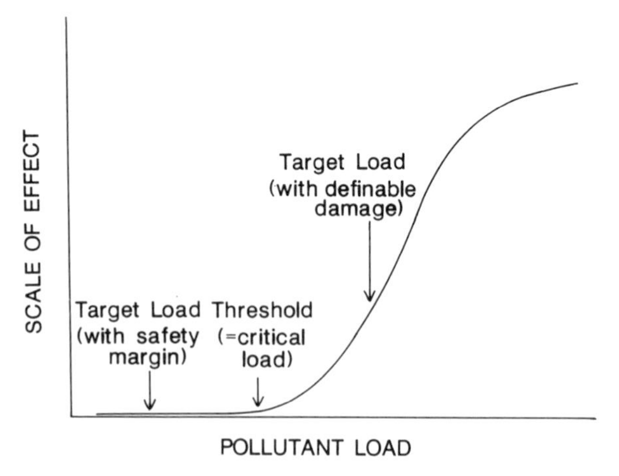

Figure 4.1 The critical loads concept (Nilsson and Grennfelt, 1988).

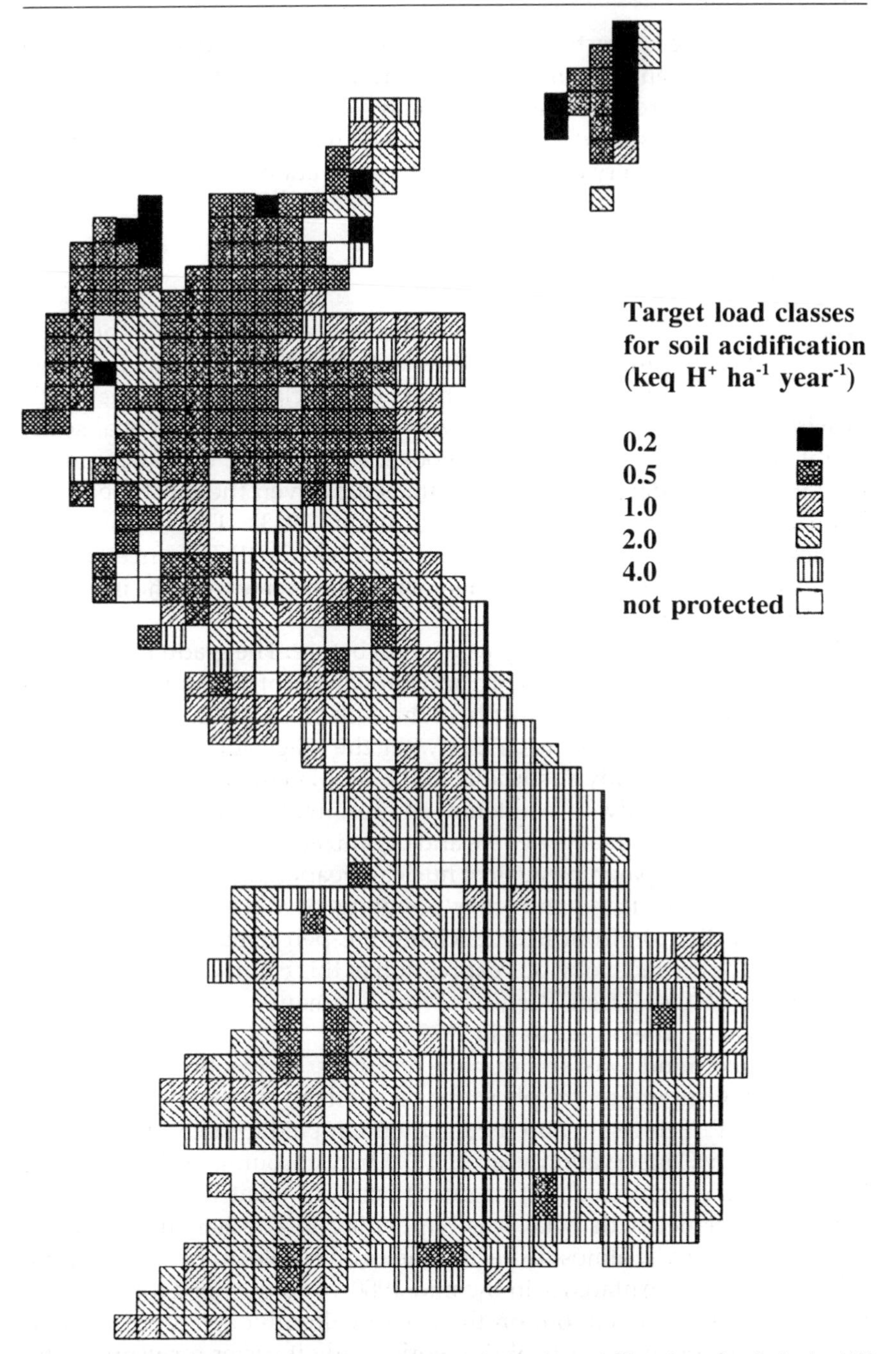

Figure 4.2 Provisional map of the target loads for the acidification of soils in the UK for the year 2005 (DOE, 1991). (Note that 8% of the country will not be protected by the planned reduction in emissions by 2005.)

whole issue of 'recovery time-scales' which are recognized by a significant fraction of the scientific community as crucial. Is it the *political judgement* – in choosing to disregard a factor that may weaken the policy's effectiveness – rather than *scientific judgement*, which will be subject to criticism if time-scales for recovery extend beyond a generation?

4.3 THE WIDER CONTEXT

Whatever the novelty of acid rain, many perceived it to be only the latest of an ever-increasing number of threats to the environment. For most observers, the radioactive fallout from the nuclear tests of the 1950s provided the first insights into how the atmosphere, on a global scale, might be polluted. Rachel Carson's books, *Silent Spring*, was a warning of how irreversible damage might be inflicted by even the most 'peaceful' application of scientific knowledge (Carson, 1963). The discovery of pesticide residues and polychlorinated biphenyls (PCBs) in Antarctic penguins emphasized that no man's backyard was free from the insidious threat (Sheail, 1985).

Perceptions were coloured by what went before. The gradual discovery of the nature of radioactive fallout, and knowledge of 'the slow circulation of strontium-90 from the stratosphere through grass and cows to the milk American children drank each day', set the stage for Rachel Carson. Americans were already suspicious of the utopian dreams of technology and more than ready to condemn environmental 'poisons' (Dunlap, 1981). Very often, the same people and organizations that aroused public conscience by campaigning against nuclear weapons and power stations, and against the pesticides and other 'polluting' products of 'big' industry, were also at the forefront in publicizing the perceived dangers of acid rain. Perhaps the most obvious expression of this common cause was the emergence of the Green Party in West Germany, and its success in winning seats in the Bundestag in 1982.

The public mind was accordingly receptive by the time of the first Stockholm Conference to pleas for 'something to be done for the environment'. Television and media attention paid to such initiatives as the European Conservation year of 1970 had encouraged the public to look for connections (real or imagined) between what was done in one part of the planet with what was happening in another. It was called '*ecology*'. The public almost expected the ozone 'hole' to be found in the stratosphere over Antarctica in the mid-1980s, and the need for controls, by international convention, on the use of chlorofluorocarbons (CFCs). The affluence of some parts of the world made it easier for users of the banned products to accept the costs of using 'environmentally-friendly' products.

A consequence of the 'Green politics' of the early 1980s was the increasing involvement of scientists in public debate. It was not enough for scientists to discover and describe threats to the environment: they were expected to prescribe remedies. Those working on acid rain frequently went straight from the conference platform of learned societies to media studios as the Scandinavian 'problem' became Europe's dilemma. There, they might be confronted with evidence that also purported to be unbiased and scientific, collected by pressure groups. In these and other ways, the difference between scientific hypothesis and 'advocacy for action' became increasingly blurred.

While all scientists seek knowledge, their primary motivation is a complex mosaic of perspectives. In the case of those engaged in environmental research, it may include a desire to improve environmental quality, secure research funds, preserve a status quo, and, perhaps self-aggrandisement. Often it is an ill-defined combination of all of these and more. The desire to be 'the first to the truth' creates intense competition and perhaps greater productivity, but it also leads to distortion. Scientists may consciously look for evidence to support their own position, or challenge that of another. Research motivated by such vested interests soon became devalued in the Acid Rain debate. The most significant achievements come from a self-critical viewpoint, supported by peer review. The search for objectivity and consensus must encompass a full range of views.

4.3.1 REMOVING UNCERTAINTIES

Assurance is sought from scientists on two points. Is a close enough watch being kept on the environment to ensure that any new or more serious form of pollution was quickly identified? Second, are there effective procedures for determining the best response to any perceived threat?

Pollution might be a global phenomenon but its incidence and impact varies considerably. Its 'randomness' was highlighted by the fallout from the accident at the Chernobyl nuclear power station in 1986. The Natural Environment Research Council's (NERC) Institute of Terrestrial Ecology (ITE), using staff from its widely-dispersed stations in the UK, collected vegetation samples from 500 sites over 10–15 days after the accident (Chapter 11, pp. 164–71). The distribution of ^{137}Caesium from the fallout material confirmed that areas in the north and west were most heavily contaminated. Heavy rain had fallen during the period when the air mass containing radioactive material was passing over those parts of the UK (Fig. 4.3) (NERC, 1987). These results were unexpected and proved to be extremely controversial. As in the early years of the 'Acid Rain' debate,

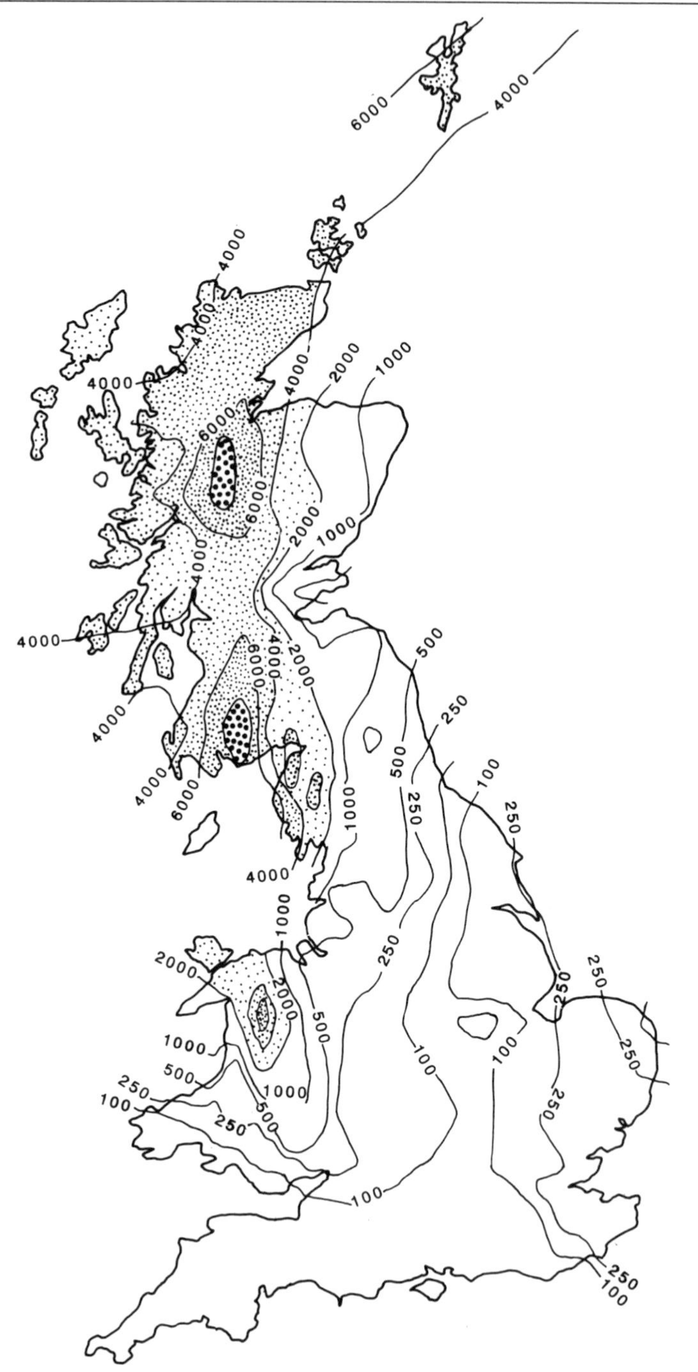

Figure 4.3 Distribution of Chernobyl fallout. ^{137}Caesium concentration in vegetation in the UK in May 1986 (Bq Kg^{-1} d.w.)

the political process had difficulty in coping with scientific evidence which challenged the status quo of perceived wisdom.

As NERC stressed, in its many submissions to Government on fighting pollution, it was quite impracticable to mount a comprehensive monitoring system, or predict the effects of every actual or potential form of pollution. Earlier mathematical models, based mainly on data from lowland sites, had predicted a fairly rapid decline in levels of radioactivity in vegetation and grazing animals. Observations by ITE showed that radioactive caesium was much more persistent in upland pastures because of the nature of the soil and the types of plant growing in these areas. Such variation in response to the same pollutants emphasized the difficulty of deciding priorities for monitoring when and where pollution might occur, and for studying how and why the same pollutant might have such varied effect on different ecosystems.

In making judgement as to what priority should be given to different pollutants, whether by the scientist or administrator, a balance has to be struck between the costs of regulating, or eliminating, the source of emission against the benefits to society as a whole (Ashby, 1978). In order to quantify the benefits there has to be a measure of agreement within and between disciplines so as to provide the basis on which politicians can make informed judgements.

Environmental scientists must agree on the mechanism, determine the dose–response relationship, and reach a consensus view on the magnitude of the effect at existing pollutant levels. The process required to achieve that position falls into several phases: observational – hypothesis formulation – experimental testing – conceptual modelling. In air pollution debates, it proved to be a drawn-out process, as vested interests in industry, pressure groups, and even academia, took up positions at the extremes of scientific uncertainty (Fig. 4.4). As a result, the public was presented, for long periods of the Acid Rain debate, with a confused view of the 'perceived risk'.

It was the role of the socioeconomist to assess the significance of the response in monetary terms. There are also several phases to this process: extract dose–response relationships – assign monetary values to environmental factor – calculate monetary value of damage. Resolution of these procedures was never satisfactorily achieved in the acid rain debate, even after a near-consensus on cause–effect relationships had emerged among environmental scientists. Major problems still arise when placing an economic value on environmental 'damage', such as the loss of amenity owing to acidification of a lake in Scandinavia and death of trees in the Black Forest, or damage to the cultural heritage.

The politician must decide the weight to be given to environmental impacts against the cost of emission controls. Government attitudes have generally followed a set pattern: ignore the problem – diplomatic

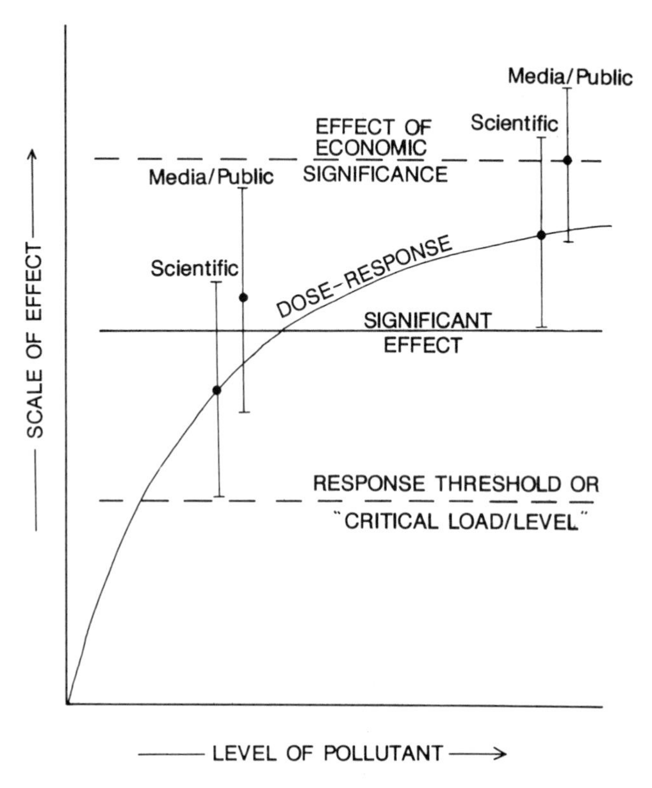

Figure 4.4 Perceptions and dose–response relationships for air pollutants (see also Ashby, 1978).

recognition with no commitment to action – support for lengthy research projects with minimal investment – announcement of controls but delay in implementation – high-profile response to international pressures to take action. This was the sequence of political events in most European countries faced with pressure to implement controls over acid rain – the main difference was that each country moved through the phases at different rates. Those in front were encouraged to, and in some instances had to, use their lead to cajole those behind. Parallels may be discerned between the policy of 'maximum visible movement and minimal action' that dominated the latter years of the acid rain controversy in the UK, and the more recent response made by governments to the debate on how carbon dioxide emissions might be curbed to protect the global climate.

What assistance might government give, whether at a national, continental or international scale, to ensure that 'the full range of views' is brought into the public reckoning? There is always a concern that official inquiries might be unduly influenced by executive or political considerations. In 1970, a *standing* Royal Commission on Environmental Pollution was appointed in the UK. Under the chairmanship of a distinguished scientist, and largely composed of academic and professional figures, the Commission was required to report on topics of its own choosing, having taken evidence from whomsoever it pleased. The focus, yet flexibility, of such an approach was evident, for example, in its Ninth Report published in 1983 on 'Lead in the Environment' (Royal Commission, 1983).

The 16 Commissioners had originally intended to conduct another general review of environmental pollution, along the lines of one made 10 years earlier in 1974. As the investigations proceeded, they decided to concentrate on the topic of lead pollution. The other issues, while important and in some cases urgent, did not give rise to nearly the same degree of public anxiety and debate; they could they felt, be saved for the next report. Despite denials from industry, it was becoming clear that lead from petrol might add significantly to the burden of the human body. The allegation that such emissions might damage health or reduce intelligence in children was much harder to appraise.

For those submitting evidence to the Royal Commission, there was the valuable opportunity of having their expertise and experience scrutinized by so prestigious a body, and of having at least a selection of material published more widely. For the purposes of the lead inquiry, it was a chance for NERC to highlight the considerable effort that had been invested over the previous decade in studying the nature and origins of lead and other elements in the UK. Largely carried out by universities, the NERC-funded research had been able to distinguish the contributions of crustal-derived materials from anthropogenic sources. A substantial proportion of the concentrations found in recent sediments of rivers, lakes and estuaries might be derived ultimately from petrol. Clearly more research was required to determine the character and significance of the more subtle effects of lead on different organisms and the physical environment but NERC, in its evidence, was in no doubt that controls over vehicle emissions should be introduced as soon as practicable.

The Royal Commission agreed. Having looked at the issues from many angles, the Commission could find no compelling arguments for retaining the use of leaded petrol. It should be possible for all new petrol-engined vehicles to be running on unleaded petrol by 1990 at the latest. The Government announced its decision to eliminate lead from petrol on the afternoon of the day the Commission's report was published. A Commissioner, the Earl of Cranbrook, later commented on how the Government had never responded so rapidly to a report of the Royal Commission. Scientists were speaking the language of politics (Cranbrook, 1991). In fact, it was concern for another form of pollution, namely NO_x and hydrocarbon emissions, that finally caused Government actively to promote lead-free petrol. Lead in exhausts would reduce the efficiency of catalytic converters. The UK has had the quickest take-up of lead-free petrol of any European country. It has also been among the first to recognize a new air pollution problem, namely increased levels of benzene in urban air.

4.3.2 THE PRECAUTIONARY PRINCIPLE

The scientific approach to problem-solving founded on the testing of hypotheses is very time-consuming. It was frequently overtaken by events in the acid rain debate. The hypothesis that acid lakes were caused by acid deposition was initially based on field data. As several distinguished ecologists pointed out, an hypothesis must be specific and subject to Koch's postulates, namely that field observations must be reproducible in both field and laboratory tests. It took nearly a decade before the numerous hypotheses formulated to explain forest decline could be subjected to the appropriate scientific tests.

Some countries were not prepared to wait that long. Politicians in West Germany cut through the 'uncertainty' by adopting a policy of 'precaution', which became in turn the basis for subsequent legislation in the European Community and on an international scale for controlling sulphur dioxide emissions. Such an approach caused considerable concern to industry, used to having 'quantifiable risks'. As the electricity industry in the UK became painfully aware, there was no limit to the insurance cover that might be sought against the possibilities of damage to the environment, or the costs of the premiums that might be paid.

The pressures to take such insurance cover have become more intense. There was a 20-year interval between the original hypothesis that lead from automobiles accumulated in humans, causing lower intelligence in children, and the decision to remove lead from petrol in the early 1980s. There was a similar gap between the hypothesis that long-range transport of sulphur emissions from western Europe caused acidification in Scandinavia, and agreement on emission-control legislation in Europe and North America in the mid-1980s. There was a 5-year interval between publication of research showing the development of ozone depletion over Antarctica and the agreement to take international action against CFCs in the Montreal Protocol in 1989. Demands for international action to limit carbon dioxide emissions preceded the scientific consensus achieved by the Intergovernmental Panel on Climate Change, published in late 1990 (Houghton *et al.*, 1990).

While time-scales have become shorter, the geographical scale has become global. If the focus of the Stockholm Conference of 1972 was on the trans-boundary movement of pollution at the continental scale, the challenge for the Brundtland Commission, 15 years later, was to draw up a blueprint for sustainable development for the planet Earth (World Commission, 1987). The elements of the precautionary principle, the polluter pays, and the definition of scientific and administrative goals, had to be addressed in a wholly international context. To achieve that, a broad scientific basis of understanding backed by powerful environmental lobbies will be required in both the industrialized and developing regions of the world. How far will the expertise and experience of the former be relevant to the socioeconomic and cultural aspirations of the latter? In that context, it was particularly appropriate that Brazil was chosen as the venue for the UN Conference on Environment and Development in 1992.

4.4 THE WAY FORWARD

Air pollution will continue to be at the top of the agenda for 'sustainable' environmental management if only because of its bearing on climate. The difficulties in matching science and policy that characterized the acid rain issue are still very much alive in the context of greenhouse gas emissions and critical loads for nitrogen compounds. On climate change, the scientific community has already made a significant contribution towards a consensus though the Intergovernmental Panel on Climate Change. However, there was again a blurring of 'scientific hypothesis' and 'call for action' – the IPCC calculation that an immediate 60% reduction in carbon dioxide emissions would be needed to stabilize atmospheric concentrations was widely reported as a call for reduction. The significant strides made to integrate economics with environmental science (Pearce *et al.*, 1989) will be a key element in developing financial incentives to curb pollution. This approach has found favour in particular with political interests looking for ways in which indirect taxation can replace income tax. Nevertheless, there are already signs that national policies are being formulated on the basis of 'visible movement with minimum penalty'.

The interactive process between the sciences and legislation is becoming clearer, as reflected in the UK's Environmental Strategy, *This Common Inheritance* (Secretaries of State *et al.*, 1990). For air pollution, the iterative process of policy formulation and evaluation based on the Critical Loads Concept will underpin international negotiations. However, political judgements based on the 'precautionary principle' must not be seen as a substitute for science-based policies. This emphasizes the need for a long-term commitment to support an adequate research base on which to develop robust environmental quality objectives.

So, not only are the temporal and spatial scales of the environmental issues becoming broader but they are becoming increasingly interdisciplinary in character. If anything, the complex nature of environmental interactions which scientific study is revealing, render the task of reconciling science and practical policy even more difficult. The scientific complexity of the acid rain debates and the unforeseen environmental consequences of large-scale installation of flue-gas desulphurization, emphasized the need for Integrated Pollution Control (IPC) in the UK. However, IPC relies on it being possible to compare and evaluate environmental impacts of quite different forms, and thus compounds scientific uncertainty with the subjectivity of value judgements. The diverse elements of regulation between government departments makes the obvious solution – the formulation of an effective Environmental Protection Agency – difficult to achieve.

In the fields of air pollution, an important goal for the UN Conference in Brazil in 1992 was an international agreement to limit carbon dioxide emissions. This attempt was made some 30 years after the signing of the Nuclear Test Ban Treaty in 1963, arguably the first occasion when scientific and public concern for the fate of the natural environment made an impact on political thinking at the global scale. The conclusion to be drawn from the Acid Rain Debate is that a sound research base must be in place before the politician and industry can make an informal response.

ACKNOWLEDGEMENTS

The authors gratefully acknowledge the helpful comments made by L.W. Blank, D.J.A. Brown, A.J. Crane, G.D. Howells and R. Wilson.

REFERENCES

Anon (1983) *Forest Damage/Air Pollution Research*, First Report of the Advisory Board of the West German State and Federal Governments, Kernforschungs-centrum, Karlsruhe.

Ashby, E. (1978) *Reconciling Man with the Environment*, Oxford University Press, Oxford.

Braekke, F.H. (1976) *Impact of Acid Precipitation on Forest and Freshwater Ecosystems in Norway*, SNSF Phase 1 Project Report, NISK, Norway.

Bull, K.R. (1991) The critical loads/levels approach to gaseous pollutant emission control. *Environ. Poll.*, **69**, 105–23.

Carson, R. (1963) *Silent Spring*, Hamish Hamilton, London.

Central Electricity Generating Board (1987) Acid rain. *CEGB Res.*, **20**, 1–64.

Central Electricity Generating Board (1981) *Submission to the Commission on Energy and the Environment*, CEGB, London.

Chester, P.F. (1986) *Acid Lakes in Scandinavia – The Evolution of Understanding*, TPRD/L/PFC/010/R86, CEGB, Leatherhead.

Clarke, A.J. and Spurr, G. (1976) Routine sulphur dioxide surveys around large modern power stations. *Atmos. Environ.*, **10**, 265–8.

Cranbrook, Earl of (1991) Ecological research and Government, in *The Ecology of Temperate Cereal Fields* (eds L.G. Firbank, N. Carter, J.F. Darbyshire and G.R. Potts), Blackwell Scientific Publications, Oxford, pp. 27–8.

Department of the Environment (1983) *Acid Deposition in the United Kingdom*. UK Review Group on Acid Rain, DOE, London.

Department of the Environment (1988) *The Effects of Acid Deposition on the Terrestrial Environment in the UK*. HMSO, London.

Department of the Environment (May 1991) *Acid Rain – Critical and Target Loads Maps for the United Kingdom*, Air Quality Division.

Dunlap, T. (1981) *DT – Scientist, Citizens and Public Policy*, Princeton, New Jersey, University Press.
Elsworth, S. (1984) *Acid Rain*, Pluto Press, London.
Fry, G.L.A. and Cooke, A. (1984) *Acid Deposition and its Implications for Nature Conservation in Britain*, Nature Conservancy Council, Peterborough.
Hallbacken, L. and Tamm, C.O. (1986) Changes in soil acidity from 1927 to 1982–84 in a forest area in SW Sweden. *Scand. J. Forest Res.*, **1**, 219.
Houghton, J.T., Jenkins, G.J., and Ephraums, J.J. (1990) *Climate Change – The IPCC Scientific Assessment*, University Press, Cambridge.
Howells, G. (1990) *Acid Rain and Acid Waters*, Ellis Horwood, London.
Innes, J.L., Boswell, R.C., Binns, W.O. *et al.*, (1986) *Forest Health and Air Pollution*, Forestry Commission R & D Paper 150.
Johnson, D.W. and Cole, D.W. (1980) Anion mobility in soils: relevance to nutrient transport from forest ecosystems. *Environ. Int.*, **3**, 79.
Natural Environment Research Council (1987) *Report for 1986/87*, NERC, Swindon.
Nilsson, I. and Grennfelt, P. (1988) *Critical Loads for Sulphur and Nitrogen*, Nordic Council of Ministers, Copenhagen.
Oden, S. (1968) *The Acidification of Air and Precipitation and its Consequences on the Natural Environment*, Swedish Natural Research Council, Ecology Committee.
Organization for Economic Cooperation and Development (1977) *The OECD Programme on Long Range Transport of Air Pollutants, Measurements and Findings*, OECD, Paris.
Parliamentary Papers, (1983–84), HC 446, Commons Environment Committee, *Fourth Report, Acid Rain.*
Pearce, D., Markandya, A. and Barbier, E.B. (1989) *Blueprint for a Green Economy*, Earthscan, London.
Pearce, F. (1987) *Acid Rain*, Penguin, London.
Roberts, T.M., Skeffington, R.A. and Blank, L.W. (1989) Causes of type I spruce decline in Europe. *Forestry*, **62**, 179–222.
Rosenqvist, I. (1978) Acid precipitation and other possible sources for acidification of rivers and lakes. *Sci. Total Environ.*, **10**, 271–2.
Royal Commission on Environmental Pollution (1983) *Lead and the Environment*, Cmnd 8852, HMSO, London.
Royal Commission on Environmental Pollution (1984) *Tackling Pollution – Experience and Prospects*, Cmnd 9149, HMSO, London.
Royal Ministry of Foreign Affairs and Royal Ministry of Agriculture (1971) *Air Pollution across National Boundaries*, Stockholm. Main report and supporting studies.
Secretaries of State for Environment, Trade and Industry *et al.* (1990) *This Common Inheritance – Britain's Environmental Strategy*, Cmnd 1200, HMSO, London.
Sheail, J. (1985) *Pesticides and Nature Conservation: the British Experience, 1950–75*, Clarendon Press, Oxford.
Sheail, J. (1991) *Power in Trust: the Environmental History of the Central Electricity Generating Board*, Clarendon Press, Oxford.
Ulrich, B., Mayer, R. and Khanna, P.K. (1980) Chemical changes due to acid precipitation in a loess-derived soil in central Europe. *Soil Sci.*, **130**, 193–9.
World Commission on Environment and Development (1987) *Our Common Future*, Oxford University Press, Oxford and New York.

Case study: the history and ethics of clean air 5

Peter Brimblecombe and Frances M. Nicholas

Peter Brimblecombe is Reader in Atmospheric Chemistry at the University of East Anglia. He is the author of Air Composition and Chemistry *(Cambridge University Press, 1986) and* The Big Smoke: A History of Air Pollution in London since Medieval Times *(Methuen, 1987).*

Frances Nicholas is a Senior Research Fellow in the School of Environmental Sciences at the University of East Anglia; she is a graduate in philosophy and theology.

5.1 INTRODUCTION

Attitudes to the environment have obviously changed in various ways over time. There has been considerable interest in the ecological awareness found among tribal communities as an outcome to their close contact with natural surroundings (see Chapter 2) but the environmental thinking of literature or early industrial societies has not been widely studied. There are exceptions; for example, White (1967, 1973) and Tuan (1975) have written about the Judeo-Christian and Chinese approaches to nature.

We do not intend to cover this ground again. Our aim is to examine certain aspects of the history of air pollution that seem to illustrate the way in which ethical ideas may have been linked to the perception and control of air pollution from the earliest time. We shall focus mainly on the

Environmental Dilemmas Ethics and decisions
Edited by R.J. Berry
Published in 1992 by Chapman & Hall, London. ISBN 0 412 39800 1

question: what made people recognize smoke as a problem? This will involve looking at responses to air pollution which arose through concern over its effects on health or on our neighbours' properties, at responses influenced by politics or religion, or by attitudes towards cities, and at modern environmental responses. We also consider some of the ways in which people attempted to deal with the problems of air pollution and why so much of the nineteenth century smoke abatement legislation was not successful.

5.2 EARLY VIEWS OF AIR AND HEALTH

The responses of early communities to air pollution most frequently relate to health. Even in ancient times doctors were opposed to human exposure to air pollutants. Strange smells and odours were linked to disease. The effects of this kind of pollution could sometimes be quite dramatic. The *Victory Stela of King Pi (ankhy)* records the Nubian siege of Hermopolis in Egypt c. 734 BC. Although the meaning is not entirely clear, military resistence collapsed when the town became too foul to be habitable (Lichtheim, 1980). 'Days passed, and Un was a stench to the nose, for lack of air to breathe. Then Un threw itself on its belly, to plead before the king.'

Classical writings about a medical awareness of the relationship between environment and health are frequent. The Hippocratic corpus contains a book *Air, Water and Places* that deals with environmental influences on disease. On a practical level there is an indication that doctors were concerned about smoke in the air. The Roman writer, Lucius Annaeus Seneca, suffered lifelong ill health and was frequently advised by his physician to leave Rome. Seneca wrote in a letter to Lucilius (*Epistulae Morales CIV*) in the year AD 61 that no sooner did he leave Rome's oppressive fumes and culinary odours than he felt better.

Arab and Islamic physicians preserved this classical learning through the Dark Ages. The Hippocratic writings were translated and widely used by Muslim physicians in the ninth century. This background contributed to the production of *Tacuini Sanitatis* or *Taqwim al–Sihhah* of Abu'l-Hassan al-Mukhtar Ibn Butlan of Baghdad two centuries later. This important work on public health emphasizes the importance of clean air and other environmental factors (Hamarneh, 1989).

Medieval European doctors generally held the view that disease was related to the air and adopted a miasmatic approach to disease aetiology; thus acrid smokes or foul odours were believed to be harmful in densely populated medieval cities. In London the smoke from coal burning was regarded as such a unique and serious risk to health that a commission was set up to investigate the problem in 1285, and was convened again in 1288 with firm instructions to find a solution. Another meeting in 1306 issued a proclamation banning the use of coal, although further legal notices issued only 2 weeks later suggest that the proclamation was largely ignored (Brimblecombe, 1987a).

This preoccupation with health has persisted. Diederiks and Jeurgens (1990) said of pollution control in nineteenth-century Leyden that, although concerned about pollution, people worried only about its effects on their health and not about the environment. There are clearly many parallels to this attitude today, with the issue of human health taking precedence in much of our effort at improving environmental quality.

5.3 OF NEIGHBOURS AND WALLS

Another thread in the development of our attitudes towards environmentally appropriate behaviour seems to be related to what would be expected of a neighbour. Old Testament teachings would have us love our neighbour (Lev. 19.18). Assyrian and Babylonian legal texts include details about the responsibilities of neighbours but little that can be seen as direct environmental awareness, except a concern for water that might be expected in an arid region. Aristotle in *Athenaion Politeia* describes the role of the *astynomoi* (controllers of the town) who were concerned with the approved discharge of water into the streets, the drains and gutters, etc. (Brimblecombe, 1987b).

Ancient cities were small and overcrowded. Under such conditions smoke, emitted at a low level from small forges or hearths, must frequently have drifted onto neighbouring houses.

Roman law gives us our first clear view of coherent and perhaps even workable environmental legislation. Although Roman law has no equivalent to the modern concept of nuisance it was aware of the 'wilful wanton interference with the enjoyment of servitude' (Buckland and McNair, 1952). The *servitudes* were sets of necessary rights of great antiquity that gave right of way to cross the land or to have access to water. There were also urban servitudes concerned with problems surrounding party walls or gutters. The following case appears to be of particular relevance: 'Aristo . . . did not think smoke could legitimately be allowed to penetrate from a cheese factory into buildings higher up in the road . . . no more can you throw water or anything else from the building higher up on those lower down . . . smoke is just the same as water.' However, the rules of abatement of nuisance, which would allow the injured person to put the matter right without legal process, did not develop formally under Roman law. The injured party had to take the accused to court. Nevertheless there is some indication that smoky industries were obliged to locate themselves on the outside of the city (Douglas and Frank, 1972), probably to keep smoke away from the wealthier suburbs. Mamane (1987) has drawn attention to parallel requirements under early Hebrew law.

Similar practices of zoning were also found in cities of medieval Europe (Brimblecombe, 1987b). Apart from these, there were relatively few methods of air pollution control available to civic administrators. There were attempts to ban or at least limit the use of coal and one group of blacksmiths in thirteenth-century London tried to restrict the times when their furnaces were used. An example of good neighbourly behaviour perhaps, but also possibly a reaction to the general dislike of the noise and pollution created by blacksmiths (Brimblecombe, 1987a).

This is well illustrated by a fourteenth-century case brought before the London Assize of Nuisance:

> Thomas Yonge and Alice his wife complain . . . the chimney is lower by 12 feet than it should be and the blows of the sledge-hammers when the great pieces of iron called 'Osmond' are being wrought into 'breastplates', 'quysers', 'jambers' and other pieces of armour, shake the stone and eastern party-walls of the house so that they were in danger of collapsing . . . and spoil the wine and ale in their cellar, and the stench of the smoke from the sea-coal used in the forge, penetrates their hall and chambers, so that whereas formerly they could let the premises for 10 marks (134 s) a year, they are now worth only 40 s.'

The defence offered by the armourers in this case is similar to that which would have been offered by many small industries in more recent times. They claimed to be honest tradesmen free to carry out their trade anywhere in the city, adapting their premises to suit their work. They rejected the complaint of nuisance on the basis that the forge they used had long been on the site and the rooms affected by smoke had been built more recently.

This case illustrates a clear awareness of economic loss imposed by smoke. Of course the complaint of air pollution may have been a method of trying to get rid of a noisy neighbour. Indeed a related incident is attributed to the painter Sandro Botticelli. He was deafened by noisy neighbours whose work shook his whole house. The neighbours insisted that they could do what they wished in their own home, so Botticelli balanced an enormous boulder atop his house which threatened to fall on their's at the least vibration; he claimed that he was free to do as he wished in his own home (Vasari, 1965).

5.4 EARLY POLITICAL INTEREST IN THE ENVIRONMENT

Attempts at smoke abatement on the part of the early municipal government in London are surprising, in view of the poor quality of the urban environment in medieval times, with dung and rubbish frequently lying about in the streets.

However the administrators were not primarily concerned about the environment in the modern sense, but with more immediate practical reasons. Untidy streets were frequently cleaned before meetings of parliament to encourage as many wealthy people as possible to come from their country retreats into the city. Some medieval English documents relating to air pollution stress that smoke would offend visiting nobles and prelates. It is a characteristic of visitors to cities, evident even today, that they are more likely to notice air pollution than are the people who have accustomed themselves to a polluted environment by living perpetually within it.

The strength of the medieval administrative reaction against coal smoke in southern England may have been aided by the fact that control policy could be directed against small groups of people who were already disliked. These groups included the kinds of businesspersons and petty industrialists who could influence the price of goods in London. The butchers, bakers and brewers not only fouled the air and the waterways, but were also the subject of many complaints about pricing. The limeburners, often blamed for early air pollution in London, were, in addition, accused of unfairly raising prices. The fact that many colliers and some limeburners would have been 'foreigners' from the north of Britain cannot have enhanced their popularity.

A particularly interesting case of discrimination was one brought against the Knights Templars in 1306 by Henry Lacy, the Earl of Lincoln. The brethren of White Friars had complained of the stench from the river. The Earl accused the Knights Templars of blocking the river Fleet by constructing a watermill. In fact, the offensive garbage was tanning waste and offal discharged by the butchers and leatherworkers upstream at Smithfield market. Despite this the Knights Templars, who owned the mill at the mouth of the river Fleet, were blamed. No doubt their mill caused some restriction to flow, but it was more probably their wealth and freedom from control by the secular authorities that made them unpopular. Persecution of the Knights Templars became intense in the years that followed and this incident can be regarded as an early case in which the stated environmental concern disguised political motives (Brimblecombe, 1987a).

Using environmental issues as an instrument for political change means there is frequently no improved environmental quality. In the case above, the waste from Smithfield market and privies along the Fleet continued to be ejected into the river. There were further problems reported in 1355 from the 'infection of the air' and the abominable stench from the discharges of latrines and tanning works. Indeed, pollution of the Fleet remained severe until the time of Charles Dickens, five centuries after the dissolution of the Knights Templars.

Political difficulties about environmental problems are not peculiar to

England. Rapid population growth in Hamburg in the sixteenth and seventeenth centuries led to considerable pollution; the local council was aware of the link between health and clean water and air, and responded by passing a large number of decrees. However, they had no effective means of enforcing these, and they were generally ignored. Nevertheless, the council's policies are significant as an expression of the contemporary idea of 'policey'; a state of good and harmonious order in society. This required a powerful administration and pious and obedient people so that social discipline could be imposed. The complexity of post-medieval town life demanded a centralized authority for control, and private exploitation of the urban environment had to be curbed. Individual citizens had a dilemma: on the one hand they were aware of the pressures that high population density imposed on the environment and the need for moralistic 'policey'; on the other, they were familiar with the benefits that accrue to the individual by ignoring individual problems, such as dumping their rubbish in the canals rather than having it carted away (Lange, 1991).

5.5 FIRE AND BRIMSTONE

In addition to political pressures for environmental control there have been religious ones. An enduring example of what has been described as a religiously-based response to a specific pollutant comes from Christian opposition to the use of coal. Chambers (1968) says that 'its sulfurous combustion products confirmed its suspected association with anti-clerical forces'. Lodge (1988) argues that the case against pollution concentrated on smoke and sulphur dioxide as the lethal ingredients on the basis of their identification with fire and brimstone in contemporary biblical pictures of hell (Gen. 19.24; Pss. 11.6; Deut. 29.23; Isa. 30.33, 34.9; Ezek. 38.22; Rev. 14.10, 19.20, 20.10, 21.8; (Lodge, 1979).

The evidence that such connections were made by early environmental thinkers is not strong. There are no clear writings about air pollution in England in the seventeenth century, although Evelyn (1661) did suggest that polluted London looked rather like a suburb of hell. Medieval alchemists like Geber and John Dastin who favoured the Lullian notion that sulphur was the source of corruption and imperfection in metals offer a little circumstantial evidence to this interpretation (Thorndike, 1958). At the beginning of the early modern period sulphur was popularly regarded as affecting health and said to produce quartan, fever and all other melancholic maladies.

However, alchemical arguments can hardly be thought of as specifically Christian, there are other traditions which viewed sulphur dioxide as an agent of disinfection and purification. A classical example of this is to be found in Ovid's *Fastes IV* or in a parallel biblical reference as: 'Magic herbs lie strewn about his tent and his home is sprinkled with sulphur to protect it . . .' (Job 18.15). It seems that the relationship between visions of 'fire and brimstone' and air pollution must remain speculative.

5.6 CLEANLINESS

The issue of cleanliness is more easily related to air pollution. Cleanliness is an important issue in many religions, and this is clear from early times until well into the present century. Regulations concerning cleanliness can be seen in the form of laws by the seventh century BC. Mamane (1987) had drawn attention to regulations in the Mishnah Laws concerning the location of offensive practices beyond and in the lee of the city walls in the second and third centuries AD. Gari (1987) argues that the Islamic belief that God favours cleanliness also led to an interest in air pollution among physicians.

Sanitary reformers of the nineteenth century can seem almost obsessive in their conviction of the relationship between 'cleanliness and godliness'; cleanliness took on moral connotations. Even early this century people such as Dr A. L. Reed could still say that 'physical dirt is closely akin to moral dirt' (Grinder, 1980).

One of the obvious signs of dirt was the smoke blackening of buildings, both indoors and out, which had drawn comment from classical times onwards. *The Odyssey* describes smoke damage to weapons hung indoors, and the blackening of temples in ancient Rome was a source of complaint. Similar complaints were to be found in medieval and early modern Europe. In England coal smoke was recognized as a cause of damage to buildings from the seventeenth century (Bowler and Brimblecombe, 1990).

For the community at large, the extra work imposed by sootfall fell largely on women, and was at the centre of complaints by them in the early modern period. These were particularly evident in smoke abatement discussions in England and North America from the late nineteenth century onwards (Jast, 1933; Grinder, 1980). Ironically, while women avoided white clothes and worked hard to ensure that their men had a clean shirt (or two) each day, the steam laundries were active in generating damaging smoke (Brimblecombe and Bowler, 1991). The concern of women about the effect of smoke penetrated into our literary heritage in novels such as Marge Piercy's *Braided Lives* (1982): 'She works hard my mother in her little house. The soot floats down day and night. Sheets on the line yellow in the acid rain of factories. Every two days the sills must be washed . . .'

If cleanliness may be next to godliness, then the relationship between humans and the pollution they create can readily be seen in religious terms.

5.7 SANITARY HYGIENE

The preoccupation with cleanliness among sanitary reformers at the end of the eighteenth century can sometimes seem obsessive to us, but the conditions they observed were truly horrible. Before this time hygiene was simply not a formal part of the notion of public health (e.g. Vigarello, 1988).

The formalization of cleanliness came as part of the sanitary legislation that typified the nineteenth century. In England the writings of Chadwick, Southwood Smith, Mayhew and others are examples of the arguments that were brought before the law makers (see Flinn, 1965). The resulting laws concerned themselves with an ever-widening range of issues relating to the health of the public. These could include anything from domestic matters such as light, ventilation and privies through mercantile and industrial issues such as adulteration of food and the discharge of waste.

Such legislation brought problems. Even the authoritarian medical officers of the Victorian period, who regarded sanitary reform as an important step forward, saw that it infringed on the rights of the individual. Indeed, it is possible to regard the reforms as simply an additional burden to be placed on the poor, with intervention in issues of health being a little more than a natural extension of mercantilism which requires a healthy workforce in order to maintain production.

There is some justification for such negative interpretations. Nineteenth century government was often preoccupied with protecting property rights, rather than improving the conditions of the poor. Nevertheless, many of the people who worked in the area of sanitary reform had a genuine desire to bring about improved conditions. The writings of medical officers and sanitary inspectors show hard-working and dedicated men; accusations that their underlying motive was to support the *laissez faire* government or rampant capitalism are overstated. They appear as people busy undertaking the very practical task of improving urban health. They were equally practical in terms of the way in which they philosophized about their work. Their ethical thinking fitted in fairly closely with the morality of Victorian society. Valuable as their ideas are, one must look elsewhere for radical or idealistic concepts of the environment, e.g. William Morris, Patrick Geddes, etc.

5.8 TOWN AND COUNTRY

Responses to air pollution have sometimes been influenced by people's attitudes to cities, although these attitudes have by no means always been negative. From Babylonian times onwards there has been a sense of conflict between town and country. Town-dwellers regarded the countryside as intellectually and culturally impoverished while towns were power-houses of liberal thought; the early modern phrase *'Stadluft machts frei'* (the air of the city makes one free) describes this well. Conversely, towns could easily be seen as a source of evil and loose morality by those who valued the countryside.

Many important Victorian writers had an intense dislike of cities and an apocalyptic mood is to be found in literature and art at the end of the nineteenth century, e.g. Richard Jefferies' *After London* (1985) and Robert Barr's *Doom of London* (1892) were stories about the destruction of the city. Poets wrote about the evil of the Victorian city with its 'dark Satanic mills'. Conan Doyle, the author of the Sherlock Holmes stories, and M.P. Shiel, attempted apocalyptic novels (*The Poison Belt* (1913) and *The Purple Cloud* (1901)) where gases annihilate the world. In the latter book the Earth's sole survivor undertakes a campaign of city-burning in order to recreate, it would seem, an Edwardian Eden (Brimblecombe, 1990).

These negative visions of the city made it easy for people to see the city as a dirty and unhealthy place, an attitude reinforced by the statistical demonstration by Graunt (1662) of the unhealthiness of urban centres. Even with this knowledge there was little attempt to improve the situation; doctors merely recommending their parents to leave town if they suffered and to go to the country. The situation was little different to that which Seneca found in ancient Rome.

5.9 ENVIRONMENTAL THOUGHT AND ACTION

At the end of the nineteenth century the desire to recreate Eden and recapture lost innocence was a recurrent theme among writers. It was acompanied by a desire to get back to nature. Such yearnings can be found in England in the writings of William Morris and Walter Crane; (Marsh, 1982; Crane *et al.*, 1987). Walter (1990) writes in some detail of the growth of environmental sensitivity in Switzerland.

The value given to the natural environment and a sense that amenity is as important as economic worth are characteristic of what we would regard as environmental thinking today. However these judgements are by no means new. The notion that industrial growth should not necessarily come at the expense of amenity can be found among quite ordinary people as early as the nineteenth century. In York of the 1850s, for example, such ideas were embodied in petitions to the local government (Brimblecombe and Bowler, 1990).

However, environmental thinking before the present century seems by and large unconnected with action. Even John Evelyn, who has been regarded as the first environmental activist, tended to be rather impractical (Brimblecombe, 1987a). Pressure to act typifies a more modern approach to the environment. Even where actions were taken on pollution problems in the past, they failed to focus on the environment as a whole. Where we see local regulations to control pollution in early modern times, they relate to a given problem in a given location (e.g. Fournel, 1813; Mieck, 1981).

Pollution problems continue to be considered independently into the twentieth century; there are still perceived to be separate problems of air pollution, noise, sewage, etc. (Melosi, 1980) despite an oft repeated commitment to an interdisciplinary approach. Thus much law is not broadly environmental in the way we might expect from a modern integrated view of the environment. Even as national laws developed there was always the problem of local interpretation (Diederiks and Jeurgens, 1990). The UK Government struggled with national uniformity in the nineteenth century when formulating the Alkali Acts, which required a central administration to avoid bias by powerful local industrialists (e.g. MacLeod, 1965). Central government sometimes had to bring pressure to bear on local councils that seemed to be avoiding the requirements of various sanitary acts of the nineteenth century.

Today the definition of environment can be almost too inclusive. As Eisenbud (1978) has put it:

> The word 'environment' has developed a new meaning since the mid-1960s. It is a word so broad in its meaning, and so inherently subjective, that it invariably reflects the interests, biases, perspectives, and motives of the user. The environment is more than the air, water, land, and presence of living things. Environmental deterioration threatens our well-being when air, water, or food become contaminated, but also when government breaks down, or our cities are destroyed by war, or free passage in the streets is interrupted, whether by muggers, choking traffic, a burst water main, or a poorly maintained road surface.

He goes on to describe how some environmentalists limit their concern to combating pollution, while others include the preservation of endangered species to national economic policy. Recently some have also included, for instance, the 'environment' of decayed inner city areas.

5.10 SMOKE ABATEMENT IN PRACTICE

So far we have largely been looking at the development of awareness of pollution and the environment, and in particular at reasons behind the recognition of smoke as a problem. However, simply recognizing a problem does not solve it. Even when people agree on what is a right action, all sorts of reasons may prevent it from being undertaken, so we should hardly expect a simple and satisfactory answer to the question:

'Why, after recognizing smoke as a problem, did people not work harder to abate it?'

If the history of air pollution control in nineteenth century Europe and North America is examined, it is clear that the injury to health, comfort and property by smoke was widely regarded as undesirable. Smoke was formally recognized as a problem as witnessed by numerous smoke abatement regulations found within sanitary legislation. Despite this, attempts to abate smoke, even in those countries with considerable legislation and well-developed local administrative structure, were only successful in isolated cases. Smoke could not simply be legislated out of existence.

The ways in which the nineteenth century Acts failed in the UK have been illustrated by Brimblecombe and Bowler's (1990) study of air pollution in late nineteenth century York. Before the 1875 Public Health Act the local administration did not pursue complaints about factory smoke in a vigorous way. They could have drawn upon various Acts (Towns Improvement Clause Act 1847, Public Health Act 1848, York Drainage, Sanitary Improvement and Foss Navigation Act 1853, Health of Towns Act 1853, Local Government Act 1858, Sanitary Act 1866 etc.), but were reluctant to place any great pressure on local industry. In serious cases, such as pollution from the iron foundry of Close, Ayre and Nicholson in 1871, it was only pressure from central government that stirred the local council into action. One reason for the failure of these early Acts in York was the absence of a clear procedure for handling smoke abatement, particularly the lack of anyone with responsibility for administering the smoke abatement regulations.

In York this changed after the passage of the Public Health Act of 1875. The council appointed its own sanitary inspector (rather than using a policeman), whose office became the focal point of attempts to reduce the smoke problem in York. Jonathan Atkinson, the inspector, was a diligent and enthusiastic officer, but the way smoke problems from individual chimneys dragged on over years seems to suggest that he faced great difficulties in lowering industrial smoke emissions (Bowler and Brimblecombe, 1989). The records suggest that there was some willingness on the part of industrialists to comply with the law, but they found compliance difficult, largely because of a lack of appropriate technology. Under pressure from the sanitary inspector, smoke abatement equipment was frequently purchased and installed, but with disappointing results. Frustrated by an inability to achieve any substantial improvement in a relatively simple way, industry tended to lose interest and resisted further pressures from a council reluctant to take the issue to court. Even under tighter laws at the end of the ninetenth century, convictions were few. Industry could easily argue that it was using the best available practice or that the smoke was not black (as required by the letter of the law); when convictions were obtained the fines were low.

Improvements in air quality came to York through the twentieth century (Brimblecombe and Bowler, 1993), but much slower than might have been expected after the installation of abatement equipment. The gradual introduction of more modern boilers, better trained stokers and an interest in fuel quality and fuel economy in the early twentieth century helped, but these were slow undramatic changes. Of course there were also improvements in the law and the Clean Air Act of 1956 is often cited, but it should be remembered that in York, for example, the Act was not fully implemented until 1990.

In the nineteenth century it was weak legislation and poor technology that hampered the elimination of smoke. However, the lack of pressure on industry to reduce smoke came from a more complex set of underlying issues. Urban wealth, prosperity and progress remained more important to most people than avoidance of environmental pollution. Lack of resources, financial and otherwise, restricted actions particularly in an era where scientific knowledge was limited and cost-benefit analysis a technique for the future. Inertia, individual greed and laziness, no doubt, all played their part, but analysis of the importance of these factors in influencing environmental actions will require more detailed study.

5.11 CONCLUSIONS

It is relatively easy to find examples of thinking on environmental problems in the past. It is more difficult to trace the way in which these early developments link to the environmentalism of the present. It is conventional to see the Romantic poets and artists of the eighteenth century as proto-environmentalists, but their philosophy often relates more to nature than to the environment. The environmentalism of our own age is necessarily rooted in that of the past, but it had taken on a singular character.

There are many important questions related to the development of modern environmental thinking, e.g. 'When was the environment first seen as coming within the realm of ethics?' and 'In what sense can it have moral status?' In traditional Western philosophy human beings have been the only proper objects of moral concern (Warren, 1983). Others have extended such ideas further to give moral consideration to the environment as a whole (Turner, 1988). Given the breadth of the modern standpoint, much needs to be done to unravel its historical development from earlier perspectives.

REFERENCES

Bowler, C. and Brimblecombe, P. (1989) The difficulties in abating smoke in late Victorian York. *Atmos. Environ.*, **24B**, 49–55.

Bowler, C. and Brimblecombe, P. (1990) *Pollution history of York and Beverley*, 2 vols, final report, March 1990, contract RK 4229.

Brimblecombe, P. (1987a) *The Big Smoke*, Methuen, London.

Brimblecombe, P. (1987b) The antiquity of smokeless zones. *Atmos. Environ.*, **21**, 2485.

Brimblecombe, P. (1990) Writing on smoke. In Bradby, H. (ed.), *Dirty Words*, Earthscan Publications, London, pp. 93–114.

Brimblecombe, P. and Bowler, C. (1990) Air Pollution in York 1850–1900, in Brimblecombe, P. and Pfister, C. (eds), *The Silent Countdown*, Springer, Berlin, pp. 182–95.

Brimblecombe, P. and Bowler, C. (1991) *Historical Evidence for the Rates of Stone Decay to the Exterior of York Minster*, final report to the National Power, May 1991.

Brimblecombe, P. and Bowler, C. (1993) Smoke and abatement activities in twentieth century York, in *Journal of Urban History: Special Issue 'The City and the Environment'* (eds C. Rosen and J. Tarr) (in press).

Buckland, W.W. and McNair, A.D. (1952) *Roman Law and Common Law*, Cambridge University Press, London.

Chambers, L.A. (1968) Classification and extent of air pollution problems, in (ed.) A.C. Stern, *Air Pollution* vol. 1, Academic Press, New York, pp. 1–21.

Crane, W. *et al* (1987) *Art and Life, and the Building and Decoration of Cities: A Series of Lectures by Members of the Arts and Crafts Exhibition*, delivered at the Fifth Exhibition of the Society, contributions from Crane, W., Cobden-Sanderson, T.J., Lethaby, W.R., Blomfield, R., Ricardo, H., published by Rivington, Percival and Co., London.

Diederiks, H. and Juergens, C. (1990) Environmental policy in 19th-Century Leyden, in *The Silent Countdown* (eds P. Brimblecombe and C. Pfister), Springer, Berlin, pp. 167–81.

Douglas, R.W. and Frank, R. (1972) *A History of Glassmaking*, Henley on Thames, London.

Eisenbud, M. (1978) *Environment, Technology and Health: Human Ecology in Historical Perspective*, Macmillan Press, London.

Evelyn, J. (1661) *Fumifugium*, London.

Flinn, M.W. (1965) Introduction to Chadwick, E. *Report on the Sanitary Condition of the Labouring Population of Great Britain*, University of Edinburgh Press, Edinburgh, Scotland.

Fournel, F. (1813) *Traite du Voisinage*, Tome II, Paris.

Gari, L. (1987) Notes on air pollution in Islamic heritage. *Hamdard Medicus*, **30**, 40–8.

Graunt, J. (1662) *Natural and Political Observations . . . Made upon the Bills of Mortality*, London.

Grinder, R.D. (1980) The battle for clean air: the smoke problem in post-civil war America, in *Pollution and Reform in American Cities 1870–1930* (ed. M.V. Melosi), University of Texas, Austin Texas, pp. 83–103.

Hamarneh, S.K. (1989) Vistas of Arabic healing arts in theory and practice. *Hamdard Medicus*, **32**, 3–54.

Jast, M. (1933) *How the Citizen takes the Air*, National Smoke Abatement Society.
Lange, N. (1990) 'Policey' and environment as a form of 'social discipline' in early modern Hamburg, in *The Silent Countdown* (eds. P. Brimblecombe and C. Pfister), Springer, Berlin, pp. 162–6.
Lichtheim, M. (1980) *Ancient Egyptian Literature, Vol. III: The Late Period*, University of California Press, Berkeley.
Lodge, J.P. (1979) An anecdotal history of air pollution. *Adv. Environ. Sci. Technol.*, **10**, 1–37.
Lodge, J.P. (1988) The big smoke. *Atmos. Environ.*, **22**, 206–7.
MacLeod, R.M. (1965) The Alkali Acts Administration 1863–1984: the emergence of the civil scientist. *Victorian Studies*, **9**, 86–112.
Mamane, Y. (1987) Air pollution control in Israel during the first and second century. *Atmos. Environ.*, **21**, 1861–3.
Marsh, J. (1982) *Back to the Land: the Pastoral Impulse in Victorian England from 1880 to 1914*, Quartet Books, London.
Melosi, M.V. (1980) Environmental crisis in the city: the relationship between industrialization and urban pollution, in *Pollution and Reform in American Cities 1870–1930* (ed. V. Melosi), University of Texas Press, Austin Texas.
Mieck, I. (1981) Umwelschutz in Preussen zur Zeit der Fruhindustrialisierung, in *Moderne Preussische Geschichte 1648–1947 Eine Anthologie*, Veroffentlichungen der Historischen Kommission zu Berlin, Bd. 52. Berlin, 2, pp. 1141–67.
Thorndike, L. (1958) *History of Magic and Environmental Science*, Columbia University Press, New York.
Tuan, Y.F. (1975) *Topophillia: A Study of Environmental Perception, Attitudes and Values*, Prentice Hall, New Jersey.
Turner, R.K. (1988) Wetland conservation: economics and ethics. In *Economics, Growth and Sustainable Environments* (D. Collard, D. Pearch and D. Ulph), Macmillan Press, London, pp. 121–59.
Vigarello, G. (1988) *Concepts of Cleanliness*, Cambridge University Press, Cambridge.
Walter, F. (1990) The evolution of environmental sensitivity 1750–1950, in *The Silent Countdown* (eds P. Brimblecombe and C. Pfister), Springer, Berlin, 231–47.
Warren, M.A. (1983) The rights of the northern world, in *Environmental Philosophy: A collection of readings* (eds R. Elliot and A. Gare), Open University Press, Milton Keynes.
White, L. (1967) The historical roots of our ecological crisis. *Science*, **155**, 1203–7.
White, L. (1973) Continuing the conversation, in *Western man and environmental ethics* (ed. I.G. Barbour), Addison Wesley, Reading, Massachusetts.

Case study: nuclear power

6

L.E.J. Roberts

Lewis Roberts is a physical chemist who worked in the Atomic Energy Research Establishment at Harwell, where he was Director from 1975 to 1986. He then became the Wolfson Professor of Environmental Risk Assessment at the University of East Anglia, Norwich, until 1991. He is the author of Nuclear Power and Public Responsibility *(Cambridge University Press, 1984) and a co-author of* Power Generation and the Environment *(Oxford University Press, 1990).*

6.1 INTRODUCTION

The generation of electricity from nuclear power currently contributes 17% of the world's electricity. The installed capacity grew from the first commercial plants in the early 1960s to reach 110 GW by 1979 and 318 GW by 1990, distributed between 26 countries. There are 96 further reactors being constructed which, if they are completed, would bring the total nuclear generating capacity to 400 GW by the year 2000. France has the highest percentage of nuclear generation (75%) and ten countries rely on nuclear stations for one-third or more of their electricity (IAEA, 1990). The overall contribution to the world's energy supply is about 6%, almost the same as the current contribution from hydropower.

Nuclear construction programmes have been cut back in recent years because of doubts about costs and safety. The two are linked, as the

Environmental Dilemmas Ethics and decisions
Edited by R.J. Berry
Published in 1992 by Chapman & Hall, London. ISBN 0 412 39800 1

additional costs of safety measures and the long and expensive procedures necessary to prove a safety case are an important factor in increasing costs. Nuclear power is capital intensive, with relatively low fuel costs. Therefore the competitive position of nuclear power depends mainly on the real discount rate assumed and on projections of future fossil fuel costs. A recent study of the lifetime costs of nuclear- and coal-fired stations, counting all costs from construction to final dismantling, showed that nuclear generation was cost competitive except in those countries that used discount rates above 8% and low prices for imported coal (Jones and Woite, 1990).

As well as a hope of lower costs, the original incentive to build nuclear stations was the prospect of a large-scale source of energy which was independent of the supply of fossil fuels. Diversity of supply is still an important strategic gain, as was emphasized by the Inspector at the recent Hinkley Point Inquiry (Barnes, 1990). He listed three advantages of a policy of diversity: (1) some security against interruptions in the supply of fossil fuels, (2) some security against volatile movements in the price of fossil fuels, and (3) security against long-term uncertainties, such as possible restrictions on burning fossil fuels because of the environmental damage that results.

It seems likely that the last point, the environmental advantages of nuclear power, will become an increasingly important factor in the arguments about its future use. These environmental advantages derive from the small quantities of fuel required and the virtual absence of atmospheric pollution. In order to generate one giga watt of electricity for a year (1 GW(e)yr), it is necessary to mine and transport only 200 tonnes of refined uranium, derived from about 100 000 tonnes of crude ore, compared with some 3.8 million tonnes of coal, or 2.2 million tonnes of oil. Land use, transport and waste disposal requirements are correspondingly low. But the absence of atmospheric pollution may be a more critical benefit.

Nuclear power stations operate on closed cycles, so that atmospheric emissions are very low. They emit no acid gases and little carbon dioxide. The environmental damage caused by the acidic gases produced by burning coal or oil and the health effects in areas of high pollution are well recognized. The possible effects on the world's climate of increasing carbon dioxide and methane concentrations in the atmosphere are currently the subject of much research. The concentration of carbon dioxide in the atmosphere is 25% higher than in pre-industrial times, mainly because of fossil fuel combustion. International negotiations are taking place aimed at stabilizing or even reducing carbon dioxide emissions (Roberts *et al.*, 1990). This will be a difficult task, given the predicted expansion of the world's population from 5 billion to 8 billion by the year 2025, an expansion that will occur mainly in developing countries which urgently need to increase their (per capita) energy use to improve their living standards.

While the most urgent target is to reduce fossil fuel use by improving the efficiency of our use of energy, environmental standards will require increasing use of non-polluting sources. These include the so-called 'renewable sources' – hydropower, solar, wind, etc. – and nuclear power. A generating station of 1 GW(e) operating for 40 years will emit some 440 million tonnes of carbon dioxide if it burns coal or 250 million tonnes plus some methane if it burns gas, while the transport and mining activities in support of a nuclear station might emit 12 million tonnes of carbon dioxide in the same time.

The environmental detriment associated with nuclear power is the spread of radioactivity. In normal operation, this is very low. The UN Scientific Committee on the Effects of Atomic Radiation (UNSCEAR) estimated that the radiation doses to the world's population from the present nuclear energy programmes add 0.03% to the background dose due to natural sources of radioactivity. However, the fear of radiation is an important element in the opposition to the nuclear power industry, which has been attacked as over-centralized, technology-dominated and secretive. Public concerns centre on the following topics (Barnes, 1990): the risks to workers and to the local population from radiation; the possibility of a serious, large-scale accident; the danger of proliferation of nuclear weapons; the need to dispose of radioactive wastes.

The administrative mechanisms currently in place to address these concerns and to control the associated risks are discussed in the following sections.

6.2 CONTROL OF IONIZING RADIATION

Ionizing radiation is the most intensively researched of any toxic agent, and the philosophy of its control the best developed. Nevertheless, the quantitative risks of the low doses of radiation which are of practical significance to the nuclear industry remain uncertain and therefore a matter of controversy, as they are small and difficult to distinguish from confounding effects.

The premier international authority is the International Commission on Radiological Protection (ICRP) which was established in 1928. The advice of the ICRP is considered in this country by the National Radiological Protection Board (NRPB). The Committee on Biological Effects of Ionizing Radiation (BEIR) of the National Academy of Sciences in the USA has also published authoritative reports. The Scientific Committee on the Effects of Atomic Radiation set up by the UN (UNSCEAR) periodically reviews the exposure of the world's population to all sources of ionizing radiation.

The quantitative measure of the energy deposited in tissue is known as the absorbed dose, measured in greys (Gy). A more useful unit is the Sievert (Sv), the 'effective dose equivalent', which is indicative of the total risk to health from any exposure to ionizing radiation (NRPB, 1987). Severe medical effects known as 'radiation sickness' occur only at doses above 2 Gy. Such large doses to the whole body are only encountered in major accidents, such as that at Chernobyl, although high doses to individual organs are used therapeutically to kill cancerous growths.

At lower doses, there is a risk of medical effects after years or decades of delay. These delayed effects occur only in a proportion of the irradiated population, with the risk of occurrence increasing with dose. (Another example of a similar 'stochastic' effect is the increased risk of lung cancer from smoking.) There are four late effects from irradiation: an increased risk of contracting cancer, which can be fatal; an increased risk of some hereditary defects; and a risk of cancer and mental handicap to unborn children. The most reliable data for cancer incidence comes from epidemiological studies of patients exposed to fairly high doses from radiotherapy, from workers in uranium mines and, most importantly (because of the numbers involved), from the Japanese bomb survivors. The evidence on hereditary effects is indirect; the children of the Japanese survivors showed no hereditary defects, and a risk figure had to be deduced from experiments with animals.

The observed incidence of excess cancer from radiation is extrapolated to a lifetime risk on the assumption that the risk at any age is increased by a factor proportional to the radiation dose received. On this basis, the ICRP conclude that the risk of fatal cancer following whole body irradiation at high dose rates is 5% per Sv for a population of all ages (Clarke, 1991b). In the absence of any definite evidence of a threshold for stochastic effects, it is assumed that any radiation dose carries some risk, although statistical evidence of excess cancer has not been seen below 0.1 Sv. But experiments on animals have indicated that the risks are lower at lower dose rates. A final figure expressing risk at the low dose rates of practical concern includes factors to allow for the assumed hereditary damage and a weighted factor for non-fatal cancer. The latest NRPB figures are 5.7% and 4.5% per Sv for a total and a working population respectively (Clarke, 1991a).

The philosophy of radiological protection has evolved on the basis of this understanding of radiation damage. Because a 'safe' threshold dose cannot be defined, regulations have to be based on the concept of an acceptable level of risk, both for radiation workers and the general public. Upper limits of radiation exposure for workers have been set so that the annual risk of delayed death from cancer at any age is not greater than the largest risks of accidental death in conventional industries, taken to be 10^{-3} per year. The NRPB recommendation is a dose limit of 15 mSv per year, where 1 mSv = 10^{-3} Sv; at this dose rate, the annual risk approaches 10^{-3} at age 70 years.

The ICRP recommended that risks to the public be reduced below those to workers by at least an order of magnitude. ICRP recommend a public limit of 1 mSv per year, averaged over 5 years in special cases. The NRPB recommendation is 0.5 mSv per year, which corresponds to a risk rising to between 10^{-4} and 10^{-5} per year beyond the age of 70 years. These risks have to be seen against the lifetime risk of dying from all causes of cancer, which is about 0.24.

The second principle of radiological protection is that all radiation doses should be reduced to levels 'As Low As Reasonably Achievable', social and economic considerations being taken into account. Application of this ALARA principle has reduced average exposures well below the maximum. The average nuclear industry worker received an occupational dose of 2.0 mSv in 1987.

All occupational and public doses from the nuclear and other industries add to a variable background of natural radiation. It has been increasingly realized over the past 10 years that natural radiation causes by far the highest health risks to the public from ionizing radiation. The average doses and range of doses from both natural and artificial causes affecting the UK population are computed by the NRPB (Hughes *et al.*, 1989). The highest average and highest individual doses are from indoor radon. Approximately 5000 people in the UK received doses above 50 mSv per year from radon, compared with two workers in the nuclear industry. The largest source of man-made radiation is from medical uses, mainly X-ray diagnostics. These UK results are typical of radiation levels worldwide (UNSCEAR, 1988; Gonzalez and Anderer, 1991).

The health risk for an individual is proportional to the total dose received, not to any particular component. Clarke and Southwood (1989) drew some interesting comparisons. The average worker in the nuclear industry receives a total dose of 4.5 mSv per year, just over half that of the average person living in Cornwall who receives 7.8 mSv per year. The members of the public most affected by nuclear discharges are a small group of people near Sellafield who eat a lot of seafood. Their total dose in 1987 was 2.9 mSv, about the same as that of a frequent air traveller. On average, nuclear discharges added less than 0.001 mSv to the annual UK dose of 2.5 mSv.

Several studies have been made of the incidence of cancer in radiation workers. No general increase has been found, except possibly for prostatic cancer (Darby *et al.*, 1985; Beral, 1990) A scheme exists for compensating the families of workers in BNFL who die of cancer if there is a reasonable chance that cancer was caused by their occupation. To date, a total of 25 compensation payments have been made.

A recent report suggested that irradiation of men working at Sellafield was associated with an increased risk of leukaemia in their offspring. These results have not been confirmed by studies of a similar cluster of leukaemia cases at Dounreay. These and other reports of leukaemia near nuclear installations are reviewed by the Committee on the Medical Effects of Radiation in the Environment. It seems that radiation levels are too low to account for the number of leukaemia cases and further studies to establish likely causes are required (Beral, 1990).

Research to establish the risks associated with low levels of radiation will doubtless continue. There are some who believe that the NRPB figures underestimate the risk, and evidence to that end was most recently considered at the Hinkley Point Inquiry (Barnes, 1990). However, no firm link has ever been established between the risk of cancer and the large variations in radiation levels that occur naturally, and this provides some reassurance that the risks have not been seriously underestimated. The policy of radiological protection seems to have provided adequate protection for workers and to have reduced exposures of the general public arising from the nuclear industry to a fraction of the variation of radiation dose across the country.

6.3 THE FUEL CYCLE AND CONSERVATION

The key stages in the provision of fuel for nuclear energy generation can be summarized as: (1) mining uranium ore and processing it to obtain a refined uranium compound, (2) chemical processing to produce fuel elements for insertion into a nuclear reactor, (3) storage of the spent fuel and transport to a fuel store or to a reprocessing plant, (4) reprocessing of the spent fuel, if required, to recover the uranium, depleted in U-235, and the plutonium for future use in fuel and (5) final storage and disposal of all radioactive wastes. The major source of radioactivity in the whole fuel cycle is the spent fuel itself. Spent fuel must be discharged into cooling ponds until the heat emission from short-lived fission products has abated, although gas or air cooling can be substituted later.

A critical decision is whether to reprocess the spent fuel, stage (4) above. The main incentive to do so is to conserve uranium supplies and reduce the need for fresh fuel. About 70% of the uranium fuel required for the second generation of gas-cooled reactors in this country, the AGRs, was recovered by reprocessing the fuel from the first generation of reactors. The recovered plutonium can also be used by adding 3% of plutonium oxide instead of U-235 in mixed-oxide fuel for Pressurized Water Reactors (PWRs); experimental fuel is being used in France and Germany. Some 20–25% of fresh uranium fuel can be saved in this way. Far larger savings can be achieved by moving from present reactor designs to 'fast' reactors (strictly, fast neutron reactors), which typically use an oxide fuel containing 25% of plutonium oxide. Fast reactors can be arranged to 'breed' fissile plutonium from non-fissile U-238, and therefore need only about 2% of the fresh uranium required by the present designs.

The known and reasonably assured resources of uranium ore available at moderate prices will be committed in about 20 years on a modest projection of world nuclear capacity of PWRs of present design, and exhausted in about 50 years (Allardice, 1990). The use of recycled fuel in PWRs would extend this time-scale by a decade or so, but a significant expansion of nuclear generation in the next century could only be sustained by gradually moving to fast reactors in 20–30 years' time. That would be technically possible. The feasibility of all stages of a fast reactor fuel cycle has been proved by the operation of the Prototype Fast Reactor at Dounreay and its associated plants, and by work in other countries. A commercial-size fast reactor is operating in France, and others are being built in Japan and the former USSR.

A reprocessing step is essential to the economy of a fast reactor fuel cycle because of the need to recycle the large amounts of plutonium used. It is also essential if the reactor fuel from present day reactors cannot be safely stored for long periods, as is the case for the British 'Magnox' reactors, which use metal fuel clad in a magnesium alloy. On the other hand, the oxide fuel used in AGRs or PWRs can be stored for long periods, and could be directly disposed of as highly active waste after several decades of cooling if the benefits of conserving uranium and building up stocks of plutonium for future fast reactors were not thought to justify the additional costs involved in reprocessing.

Reprocessing fuel has been criticized on three grounds: (1) the effluent is a major source of contamination, (2) reprocessing generates volumes of radioactive waste, and (3) separating plutonium adds to the danger of proliferation of nuclear weapons – a point considered in the section 6.5. Discharges from Sellafield, which did increase in the 1970s, have been reduced by building a series of plants to trap the radioactivity, and discharges will fall to less than 1% of their maximum by 1992. The Radioactive Waste Management Advisory Committee (RWMAC) has compared the waste arisings and impacts of early reprocessing, delayed reprocessing and no reprocessing and found them all small and comparable (RWMAC, 1991).

The House of Lords Select Committee on the European Communities (1988) also considered the case for reprocessing and concluded that Magnox fuel has to be reprocessed, on environmental grounds, and that the THORP plant at Sellafield for reprocessing oxide fuel should be completed to undertake the contracts now in hand. This plant is contracted to reprocess fuel from eight countries. However, the Committee welcomed plans for dry stores for oxide fuel as allowing a degree of flexibility in reprocessing schedules, and recommended that a critical review of the future of oxide fuel reprocessing should be carried out during this decade.

6.4 WASTE DISPOSAL

The disposal of the radioactive wastes arising from reactor operations and from reprocessing plants has come to be perceived by many sections of the public as one of the key problems facing the nuclear industry. This topic, combined with the fear induced by the word 'radiation', seems to have struck deep chords of public unease and of suspicion of technology, of government and of expert evidence. This view differs sharply from the judgement of most, if not all, of the members of the technical community who would rate the risks arising from the final act of waste disposal as the least from the whole fuel cycle rather than the largest (UNSCEAR, 1982). This technical judgement rests on three perceptions: the durable nature of the packaging made possible by the small volumes to be handled; the decrease of the radioactivity over time to levels comparable with indigenous rocks; and the essentially simple technology involved.

Wastes are characterized by their radioactive content per unit volume. Heat-generating waste (HGW) is the small volume of fission products isolated after reprocessing spent fuel. The radioactivity and associated heat emission are high enough for artificial cooling to be required for some decades. Low-level waste (LLW) is slightly radioactive rubbish which can be handled without shielding, while intermediate-level waste (ILW) requires shielding but no artificial cooling. As a rough guide, the lifetime output of a large PWR would give rise to 260 m^3 of HGW, 2400 m^3 of ILW and 33 000 m^3 of LLW. Low-level waste is compressed into steel drums. Intermediate-level wastes are commonly incorporated into concrete blocks. The heat-generating wastes are mixed with glass-forming materials and cast into blocks of glass inside steel cylinders, which can then be stored behind shielding and cooled in a current of air. Spent fuel, if it is to be treated as waste, is similar to HGW and must be kept cool, and packed in conducting material in thick metal containers before disposal.

Incorporating HGW and ILW into durable, solid forms is a most important step in achieving safe storage and disposal, as dispersion is thereafter scarcely possible. While HGW should be stored with artificial cooling for some decades to ease handling problems, both ILW and LLW can be moved to permanent disposal whenever repositories are available. However, storage is perfectly feasible so long as the stores can be maintained. The decision to dispose permanently of all categories of waste in due course rather than to store indefinitely is a political decision based on long-term environmental and ethical considerations, not on technical necessity. The principle is that the wastes should impose no burden of care or significant hazard on later generations. This policy was advocated by the Royal Commission on Environmental Pollution (1976), and reiterated since then as a moral duty in many parliamentary reports and government papers. Permanent disposal underground is now seen as the only practicable solution by every country with a nuclear programme (Roberts, 1988).

Once the waste is in a repository, radionuclides can return to the biosphere only through solution and migration in groundwater, unless the repository is disturbed. Such migration is impeded by a succession of barriers. These include engineered barriers, involving the durability of the containers and of the wasteform itself, and the properties of the filling material which surrounds the waste. In addition, the natural barriers are the resistance to water flow of the surrounding geological strata, and the eventual dilution in the biosphere. The depth at which repositories are constructed depends on the radioactive content and its half-life. Repositories in deep trenches have been used successfully in many countries for LLW and ILW of short half-life. All HGW and ILW of long half-life must be disposed in deep repositories to increase the distance from the biosphere and to minimize the possibility of disruption during long periods of time.

Very stringent criteria have been imposed by regulatory authorities before repositories will be licensed (Roberts, 1990). The criteria published by Environmental Departments in this country are among the most severe. These require proof beyond reasonable doubt that the maximum radiation dose to any individual arising from the waste in a single repository at any time in the future will be 0.1 mSv per year. This limit is one-fifth of the annual limit set for today's population, one-twentyfifth of the average natural background level and far less than the variation in background level across the country.

Presenting such a case for the indefinite future is a major intellectual challenge and extensive research efforts are being deployed in many countries. Progress in this country towards a safety case for the deep burial of ILW has been reviewed by the IAEA, by independent consultants and by the RWMAC (1991). The most comprehensive exercise on the disposal of HGW was the international PAGIS project (1988). This project estimated migration from repositories in clay, granite and salt formations and also from a sub-seabed location. Particular attention was given to the effect on the results of uncertainties in the parameters chosen.

The results of these and many other similar exercises is that the probability of any radioactivity reaching the surface from a well-chosen location is low for 10 000 years and that the maximum dose arising at longer times would be below 0.1 mSv per year. Furthermore, the residual radioactivity remaining after 10^4–10^5 years is similar to that left in the uranium mill tailings, which contain all the natural radionuclides of the uranium decay series. The highest doses in the distant future are therefore likely to arise from these heaps of mill tailings, not from the buried waste, as the mill tailings are stored on the earth's surface, while the wastes will be more than 300 metres underground (UNSCEAR, 1982). The RWMAC (1991) make another comparison. The long-lived activity in the proposed ILW repository arises from the uranium in the wastes and its natural decay products, and the amount of uranium present is about the same as in similar volumes of the surrounding rocks.

It seems then that the very strict criteria proposed by the regulatory authorities can be met. However, the key parameters in the calculations are the rate of flow and direction of flow of groundwater, which can only be established by hydrogeological measurements carried out at the depth of the repository. In this country, UK Nirex Ltd has recently been given permission to drill some bore holes at two sites, with a view to deciding which is the best for ILW disposal. A more thorough examination of underground flow patterns will have to be undertaken before a final safety case can be made.

The pace at which site investigations can be carried out has been slow because of public opposition and consequent political concern. Every proposal has been opposed by groups opposed to the nuclear power programme, who have found this issue a useful one on which to focus public fears and opposition. The shifts and changes in Government policy have not helped to instil confidence and have been criticized sharply by the Advisory Committee (RWMAC, 1986). Opposition to radioactive waste disposal has occurred in most countries with a nuclear programme, although better progress has been made in these than in the UK (PAGIS, 1988). The roots of this opposition have been the subject of much speculation and of sociological research. It is necessary to show that decisions are made after an open, honest and thorough investigation. Local concerns are broader than the question of long-term radiation doses; they include worry about transport risks, the effects on property values, local industries, farming and tourism (Kemp, 1990). These worries are no different in kind from those that would be expressed in many remote areas designated for new industrial development. But they are exacerbated in this case by the fear of radiation, and what people are expressing is the social stigma that is now associated with radioactivity. Such a situation can only be remedied by patient exposition of the case, early involvement of local authorities and sympathetic consideration of a policy of improvement in the local environment. The national consultative exercise around the document *The Way Forward* (Nirex, 1988) was a useful step, as is the deeper involvement of the Advisory Committee in the site selection process (RWMAC, 1991). It remains to be seen if these measures are sufficient to win at least a measure of public acceptance.

6.5 NON-PROLIFERATION OF NUCLEAR WEAPONS

The need to prevent the diversion of nuclear materials from civil to military use has been recognized since the early 1950s. The International Atomic Energy Agency (IAEA) began to apply its own safeguards in 1959. The major instrument in force now is the Non-Proliferation Treaty (NPT) signed in 1968. The main provisions of the NPT are:

1. Nuclear weapon states undertake not to transfer to anyone nuclear weapons or the means to manufacture them;
2. Non-nuclear weapon states undertake not to manufacture nuclear weapons;
3. Each non-nuclear weapon state undertakes to accept safeguards as negotiated with the IAEA;
4. All parties agree to facilitate the peaceful development of nuclear energy;
5. The parties agree to pursue negotiations leading to the end of the nuclear arms race and to nuclear disarmament.

By 1985, 124 states had acceeded to the NPT, including three nuclear weapon states, the former USSR, the UK and the USA. Since then, at least seven more states have signed. The IAEA is applying some safeguard provisions in other non-nuclear weapon states. There are also local treaties: Euratom, including France, has a safeguard agreement with the IAEA, and the Treaty of Tlatelolco prohibits nuclear arms in Latin America.

The essential provision of safeguards agreement is a guarantee of access for IAEA inspectors to declared nuclear sites in order to account for all nuclear material and to detect unauthorized diversion, using a variety of techniques which have developed in sophistication over time. The way this system works was reviewed by Fischer and Szasz (1985). The main strength of the IAEA is the political hazards that would be run by any state reported to be not complying with the provisions of the NPT. The IAEA has no force at its command, but may propose further inspections to sites not specified in the original agreement if there is evidence justifying such action.

The non-proliferation regime can claim to have been successful. The NPT is the most widely supported arms control agreement in history (Waldegrave, 1990). The prophesies once made of many states arming themselves with nuclear weapons by now have not been realized, and the trust between states engendered by the NPT and the IAEA safeguards must have been important contributory factors to this result, although the system is not perfect; the bombing of the Iraqi research reactor by Israel was spectacular evidence of that. Furthermore, the recent reports from UN inspectors in Iraq show that a country can plan to develop covert nuclear weapons even after signing the NPT if it decides to renege on treaty obligations.

There is also natural concern about some states which have not concluded safeguard agreements although they are signatories to the NPT, and perhaps more about states outside the NPT which are known to possess the means of building up stocks of fissile material. These include India, Pakistan, Israel, Argentina and Brazil (Waldegrave, 1990). The declared weapon states, the USA, former USSR, UK and, recently, France, have placed their civilian plants under IAEA safeguards to meet the criticism that they might otherwise be at a commercial advantage over non-weapon states such as Germany and Japan which have strong civilian nuclear construction programmes; China may soon follow suit.

The point at issue in this chapter is whether an expansion in the civilian use of nuclear energy would lead to a weakening of the non-proliferation regime. Such a result is unlikely. All the declared weapon states developed weapons before developing a civil nuclear programme and not as a result of having done so. It is simpler and cheaper to build clandestine enrichment plants to produce U-235 or small research reactors to produce Pu-239 than to divert material from large, power-producing reactors. Such diversions would have to be organized for lightly irradiated fuel, with the plutonium still largely Pu-239, the fissile material used in weapons. Fuel in modern commercial reactors is taken to high irradiation levels, and discharged with an isotopic composition of about 54% Pu-239, mixed with higher isotopes that would be difficult to separate. It is not useful weapon material, although a terrorist device might be made from it.

The possible switch in the future to a fast reactor fuel cycle would introduce one new requirement. Fissile material 'bred' from U-238 would be nearly pure Pu-239, and safeguard monitoring would have to be applied to ensure that this material is mixed with the plutonium from spent fuel before final purification (Jordan and Roberts, 1990). With that exception, the safeguard requirements would be similar to those at today's reprocessing plants.

This section has been concerned with deliberate diversion of nuclear material by states, and the international control regime designed to inhibit such activities. The protection of nuclear material against criminal attack is a matter for national security measures in all countries with nuclear programmes. These normally consist of some control on access to plants, and strict accounting for nuclear materials.

6.6 RISK OF MAJOR ACCIDENTS

The characteristic accident that can affect a nuclear reactor is overheating of the fuel leading to release of radioactive material. A reactor cannot explode like a nuclear bomb, which depends on forcing together a critical mass of nearly pure fissile material. However, fuel can overheat owing to loss of control of the neutron level or to failure of the cooling system. Cooling has to be maintained even after the reactor has been shut down, owing to the heat emitted from the radioactivity of the fission products.

The release of radioactivity is prevented by several barriers: the cladding of the fuel elements, the primary cooling circuit in which the fuel is placed, and the reactor containment. Every reactor is equipped with instrumentation designed to indicate when temperature or neutron levels exceed prescribed limits and with safety systems designed to prevent any component failure precipitating a radioactive release. The principles of redundancy, diversity and containment on which safety systems should be designed are now well established (Roberts *et al.*, 1990). However, there is always some chance that all safety systems will fail. Methods of quantifying the risks of failure have been developed to a high state of sophistication, following the pioneer work of Farmer (1967) and Rasmussen *et al.* (1975). There are two stages in all such calculations. The first is to establish by a logical mathematical method the chance of some internal failure or of some external event leading to a total plant failure and a release of radioactivity. The second stage consists of a calculation of how the released radioactivity would disperse, taking account of meteorological conditions, and of the consequent risks to workers and to the surrounding population, after allowing for possible protective action, such as sheltering or evacuation. Other consequences, such as the area of land that might be contaminated, can also be estimated.

A quantitative risk assessment is now an accepted part of the safety case that has to be made for any new nuclear reactor. But it is only part of the case. It must be accompanied by a description of the operating procedures, the standards of operator training, and the safety organization that will monitor the behaviour and maintenance of the plant. In this country, the safety case is examined by the Nuclear Installations Inspectorate (NII) of the Health and Safety Executive. A licence is required from the NII before construction can commence and again before the plant can be commissioned and operated. The NII regularly inspect operating stations and all incidents involving an unauthorized release of radioactivity, however small, have to be logged. All incidents are reported quarterly by NII (HSE, 1987).

While the details of a mathematical risk assessment are a matter for an expert Inspectorate, the principles on which a safety case is based must be clear and the basis on which judgements are made must be transparent. This is the remit of the Advisory Committee on the Safety of Nuclear Installations, which publishes an annual report. The Health and Safety Executive (1987) has also published a paper for general discussion, setting out the standards to which they work. But the most searching public examinations of the safety case for nuclear reactors that have occurred in this country are the two public inquiries that have been held on the proposals to build PWRs at Sizewell and Hinkley Point (Layfield, 1987; Barnes, 1990). In each case, the entire safety case was examined, including the standards of radiological protection, the organization and results of quantitative risk assessment, the standards and methods of the NII and the organization of emergency procedures. The evidence of objectors at all points was considered and weighed.

The final reports were works of massive authority. The Inspector at Hinkley Point concluded that the maximum tolerable risk from the operation of this PWR would be a risk of early death or fatal cancer of 10^{-5} per year for any individual member of the public and 10^{-6} per year for any sizeable number of the public. He equated these risks to those faced by someone living 1 km and 5 km from the site and was convinced that the actual risks would be lower than these figures. The Inspector at Sizewell based his recommendations on the qualitative evaluation of the competence and reliability of the engineering, of the standards of quality assurance, of the operational control and of the strength of the licensing process. He undertook an evaluation of the quantitative evidence on risk as a final check on these qualitative conclusions. He made a straightforward risk calculation on the evidence before him of the total numbers of cancer deaths that might occur in the UK from the normal operation of the Sizewell B reactor. The annual social risk was calculated to be 16×10^{-4} from normal operations and 4×10^{-4} from accidents of all sizes. The Inspector thought that this level of social risk, one death in 500 years, was 'a minute risk by any standards', and noted that making an allowance for aversion to accidents involving large consequences would not significantly alter this conclusion, as the risks were dominated by the risks of normal operation.

However, the scale of consequences is important in judging the acceptability of risks and the analyses show there remains a very small chance of an accident involving many people in the event of a large release of radioactivity coinciding with a break in the reactor containment and the worst weather conditions affecting dispersion. For example, the risk analyses reported at the Hinkley Point Inquiry included one case which might lead to between 0 and 960 early deaths and 140 to 53 000 cases of fatal cancer over 70 years, depending on weather conditions. However, the probability of the worst consequences of this case was calculated to be 2.9×10^{-10} per year, i.e. 3 in 10 000 000 000 years.

There have been two accidents to commercial reactors that have involved a release of radioactivity and risk to the public. The first was to a PWR at Three Mile Island, Pennsylvania, in March 1979. This resulted from loss of cooling water after the reactor had been shut down, caused by the misinterpretation of instrument signals in the control room by the operating team. The escape of radioactivity was small because the reactor containment remained intact. The actual release was estimated to cause between zero and five cancer cases in the area in the next 30 years, but the public alarm was considerable and caused a loss of confidence in the industry (Kemeny, 1979). The subsequent analysis of the accident sequence gave a considerable boost to the use of quantitative risk assessment and two new institutes were set up in the USA; the Nuclear Safety Analysis Center, and the Institute for Nuclear Operations (INPO). All US utilities with nuclear plants are members of INPO, as are organizations in 14 other countries. The main functions are: the evaluation of all nuclear plants; raising standards of nuclear operator training; examining any abnormal events; and communicating the lessons learned to other utilities.

Unfortunately these lessons were not learned in the former USSR. The accident to a Russian RBMK reactor at Chernobyl in April 1986 was infinitely worse. As a result of gross mismanagement of a safety test, the reactor experienced a power surge which led to fragmentation of the fuel and to a steam explosion that ruptured the containment. Up to 50% of the volatile fission products and 3–6% of the fuel were dispersed. Very high radiation doses were suffered by some 260 people on the site and 31 have died. No one outside the site received large enough doses to cause acute radiation syndrome, but more than 135 000 people were evacuated from nearby villages which were contaminated with radioactivity. The radiation doses to regional populations from the Chernobyl accident have been summarized by UNSCEAR (1988). The highest doses were received in Eastern Europe, where the average dose increased by 32% of background dose in the year following the accident. The total radiation doses to the populations of Eastern and Western Europe over the next 30 years will effectively increase by amounts equivalent to 6 months and 3 weeks of background respectively. The total world collective dose from Chernobyl is equivalent to about 20 days of average background levels, delivered over 30 years, which may give rise to between 8000 and 30 000 fatal cancer cases (Clarke, 1989). As a comparison, the weapon tests carried out in the atmosphere between 1952 and 1981 caused a much larger dispersion of radioactivity, with a world average dose equivalent to nearly 3 years of natural background, delivered over a long period of time.

The Chernobyl accident was the result of serious faults in the design of the RBMK reactors and was precipitated by gross incompetence on the part of the operating team, who disobeyed several safety instructions (Gittus *et al.*, 1988). The accident illustrated the importance of achieving world-wide acceptance of standards of design and of safety organization. Since 1986, the IAEA have taken several initiatives aimed at raising standards of reactor safety, and nuclear utilities have formed the World Association of Nuclear Operators (WANO) with offices in Paris, Moscow, Tokyo and the USA. The aims of WANO are similar to those of INPO in the USA, and include the rapid exchange of information about significant events, together with analysis of these events and the exchange of operating experiences. WANO is forming a special team to assist with the safety problems of the older reactors in Eastern Europe. It is much to be hoped that safety standards everywhere will be improved as a result of more open communication.

6.7 CONCLUSIONS

The nuclear industry can reasonably claim to have been set some of the highest safety standards applied to any industry, and to be subject to an elaborate apparatus of national and international control. Comparisons with fossil fuel burning are favourable. Radioactive pollution is very low compared with the atmospheric pollution from fossil fuel combustion, and emissions are strictly monitored and controlled. The nuclear industry was a leading pioneer in applying risk analysis to large plants, and the record of serious accidents is also good compared with that in the other large-scale energy industries (Table 6.1). Modern reactors constructed to Western standards are very safe, although there always remains a remote chance of a serious accident affecting many people. However, the safety systems required to attain these standards are complex and the details of the risk analysis are difficult to understand. It is not surprising that the nuclear industry is turning towards designs for the next generation of reactors which are simpler and rely on the laws of physics rather than on elaborate engineering to attain high safety standards (Roberts *et al.*, 1990). Unprecedented standards are being applied to radioactive waste disposal. The safe disposal of highly radioactive waste remains to be accomplished but the means to be adopted are now clear. As an ultimate disposal problem, it is simple compared with the collection and disposal of billions of tons of carbon dioxide a year in such a manner that the gas cannot return to the earth's surface, if that should ever be necessary to avoid serious climatic change.

Table 6.1 Severe accidents world-wide in the period 1969–86

Industry	No. of accidents	Installation concerned	Number of immediate fatalities per event
Coal	62	Coal mines	10–434
Oil	6	Oil platforms	6–123
	15	Refineries and tank farms	5–145
	42	During transport	5–500
Gas	24	Various fires/explosions	6–452
Hydropower	8	Dams	11–2500
Nuclear	2	TMI and Chernobyl	0–31*

* Plus 200 cases of radiation sickness at Chernobyl.
+ After Fritzsche (1989).

The attainment of high standards in plant construction and operation, radiological protection, fissile material control and waste disposal depends more than anything else on the dedication to safety of the organizations concerned and on the quality and training of the people in them. This requirement applies as much to the regulatory authorities as to the utilities themselves and is a requirement common to all high-technology industries. There remains the question how such qualitative judgements can be made. In practice, the critical occasions in the UK have been the public inquiries into proposals for new plants, which have involved a prolonged interrogation of proposers, regulators and objectors. The eventual recommendations of the Inspector are passed to the Secretary of State for decision and his decision is subject to Parliamentary scrutiny. This system has many strengths. All evidence is subject to searching cross examination and the final reports are models of accuracy and consistency. However, some improvements in procedure could be made (O'Riordan *et al.*, 1988) and the inquiries are cumbersome and repetitive. For example, the principles of radiological protection have been examined *ab initio* in four public inquiries in this country. Admittedly new evidence accrues from time to time, but the question remains as to whether some more continuous public appraisal of this and other technical questions would be advantageous.

How to satisfy the energy requirements of an expanding world population in an environmentally acceptable manner is a major dilemma. The contribution from non-polluting technologies must surely grow. No one of these will be sufficient for all needs; all reasonably economic sources will be needed. Nuclear power, as an established large-scale technology, is well placed to make a larger contribution, particularly to the energy needs of the industrialized countries, which have caused most of the pollution and which should take the lead in reducing it. However, the part nuclear power will actually play depends on the degree of trust that can be won by the industry and the organizations set up to control it.

ACKNOWLEDGEMENTS

The author is indebted to many colleagues for information and discussions and in particular to Dr Thomas Ekered of WANO, and Dr K.S.B. Rose and Mr K. Taylor of AEA Technology, Harwell.

REFERENCES

Allardice, R.H. (1990) Nuclear fuel reprocessing – the reason why. *Nucl. Engineer*, **31**, 142–5.

Barnes, M. (1990) Report. *The Hinkley Public Inquiries*, HMSO, London.

Beral, V. (1990) *Br. Med. J.*, **300**, 411–12.

Beryl, V. *et al.* (1985) Mortality of employees in the UK Atomic Energy Authority, 1946–79. *Br. Med. J.*, **291**, 440–7.

Clarke, R.H. (1991a) Radiation protection standards, in *Innovation and Environmental Risk*, Belhaven Press, London.
Clarke, R.H. (1991b) Suppl. *Radiological Protection Bulletin*, No. 119, NRPB, Chilton, Oxfordshire.
Clarke, R.H. and Southwood, T.R.E. (1989) Risks from ionizing radiation. *Nature*, **338**, 197–9.
Farmer, F.R. (1967) Reactor safety and siting: a proposed risk criterion. *Nucl. Safety*, **8**, 539–48.
Fischer, D. and Szasz, P. (1985) *Safeguarding the Atom: a Critical Appraisal*, Taylor and Francis, London.
Fritsche, A.F. (1989) Health risks of energy production. *Risk Analysis*, **9**, 565–77.
Gittus, J.H. *et al.* (1988) *UKAEA Report NOR 4200*, 2nd edn, HMSO, London.
Gonzalez, A.J. and Anderer, J. (1991) *Nucl. Energy*, **30**, 27.
Hughes, J.S., Shaw, K.B. and O'Riordan, M.C. (1989) *The Radiation Exposure of the UK Population, 1988 Preview*. Report NRPB-R227, NRPB, Chilton, Oxfordshire.
Health and Safety Executive (HSE) (1987) *The Tolerability of Risk from Nuclear Power Stations*, HMSO, London.
International Atomic Energy Agency IAEA (1990) *Power Reactor Information System*, Vienna.
Jones, P.M.S. and Woite, G. (1990) *IAEA Bull.*, **32**, 18–23.
Jordan, G.M. and Roberts, L.E.J. (1990) Environmental aspects of the fast reactor fuel cycle. *Phil. Trans. R. Soc. (Lond.)*, **A331**, 395.
Kemp, R. (1990) Why not in my backyard? *Environ. Planning A*, **22**, 1239–58.
Kemeny, J.G. (1979) *Report on the President's Commission on the Accident at TMI*, Washington DC.
Layfield, Sir Frank (1987) *Sizewell B Public Inquiry Report*, HMSO, London.
National Radiological Protection Board (1987) *Living with Radiation*, HMSO, London.
O'Riordan, T., Kemp, R. and Purdue, M. (1988) *Sizewell B: An Anatomy of the Inquiry*, Macmillan, London.
PAGIS (1988) *Report EUR 11775EN*, EEC, Luxembourg.
Radioactive Waste Management Advisory Committee (1986) *Seventh Annual Report*, [RWMAC] HMSO, London.
Radioactive Waste Management Advisory Committee (1991) *Eleventh Annual Report*, [RWMAC] HMSO, London.
Rasmussen, N. *et al.* (1975) *Reactor Safety Study*, Report WASH-1400, Nuclear Regulatory Commission, Washington DC.
Roberts, L.E.J. (1988) Radwaste: spectre or symbol? *R. Inst. Proc.*, **59**, 259–77.
Roberts, L.E.J. (1990) Radioactive waste management. *Ann. Rev. Nucl. Part. Sci.*, **40**, 79–112.
Roberts, L.E.J., Liss, P.S. and Saunders, P.A.H. (1990) *Power Generation and the Environment*, Oxford University Press, Oxford.
Royal Commission on Environmental Pollution (1976) *Sixth Report*, Cmnd 6618, HMSO, London.
United Nations Scientific Committee on the Effects of Atomic Radiation (UNSCEAR) (1982) *Report to the General Assembly*, United Nations, New York.
United Nations Scientific Committee on the Effects of Atomic Radiation (1988) *Report to the General Assembly*, United Nations, New York.
Waldegrave, W. (1990) Speech at *Fourth Non-Proliferation Treaty Review Conference*, Geneva.

Case study: agricultural plenty – more or less farming for the environment?

7

B.H. Green

Bryn Green has been on the staff of the University of London's Wye College since 1974; he is now the Sir Cyril Kleinwort Professor of Countryside Management. Previously he was a Regional Officer of the Nature Conservancy Council. He is a member of the Countryside Commission for England and formerly a member of the England Committee of the Nature Conservancy Council. He is the author of Countryside Conservation *(Unwin Hyman, 1985), joint author of the* Diversion of Land *(with C.A. Potter et al., Routledge, 1991) and the* Changing Role of the Common Agricultural Policy *(with J. Marsh et al., Belhaven, 1991).*

7.1 INTRODUCTION

Until recently farmers were widely regarded as the 'custodians of the countryside' and most still see themselves in this way. This role has been generally taken to include the handing of the land to their successors in 'good heart' (i.e. fertility), and the maintenance of the amenities of the countryside, including its richness in wildlife and scenic beauty. In other

Environmental Dilemmas Ethics and decisions
Edited by R.J. Berry
Published in 1992 by Chapman & Hall, London. ISBN 0 412 39800 1

parts of Europe, notably in France, small peasant farmers are additionally seen as part of the *'patrimoine'* and there are powerful forces for the preservation of this way of life. Similar feelings are strong in parts of the UK, where, for example, the Welsh and Scottish languages and cultures are now mainly rooted in remote hill farming communities.

The history of agriculture in Europe is by no means free of the severe environmental impacts such as erosion and desertification which presently accompany its practice in many parts of the world. In the UK the climax forest which covered most of the land now farmed had been largely cleared before the present millenium; it has been estimated that when the Domesday Book (1086) was compiled that the forest cover was 15%. (Rackham, 1986). Recent archaeological research suggests that the Neolithic and earlier Mesolithic clearances were much more extensive than was formerly supposed. The environmental impact of this first agricultural revolution was probably just as great as that of the forest clearances in the Tropics which are now causing so much concern. We know that over large parts of the uplands peat formation was initiated and that many species, including bear, beaver, wolf and perhaps also aurochs and bison, were lost. There is also evidence that medieval farming had some very damaging environmental impacts. Common grazings were often grossly overstocked, resulting in erosion, which occurred particularly on the light, sandy soils of heathland. The second major agricultural revolution of the Parliamentary Enclosures also had very damaging effects on the environment. As well as the land rationalization that converted peasant farming to capitalist agriculture, the new technologies of crop rotation and plant and animal breeding enabled large tracts of marginal lands, like heath, marsh and down to be brought into more intensive production. Species such as the great bustard and marsh harrier were lost as a result.

Overall however, it is arguable that traditional agriculture has diversified and ornamented the rural environment. The scenery, wildlife and cultural elements of the countryside have been enhanced by replacing some of the originally almost continuous temperate forest climax communities by plagioclimax grasslands, dwarf-shrub heathlands and wetlands, particularly through forest clearance, drainage and the introduction of pastoral systems of agriculture. The resulting small-scale mosaic of semi-natural and farmed and forested vegetation, with its settlement pattern of villages and small country towns, has been greatly cherished for its beauty, environmental quality and the access it provides into the countryside. It is also arguable that traditional land uses were deployed in relation to the long-term carrying capacity of the land. Thus woods were left on some of the more infertile or intractable soils, grazing marshes on washlands and thin, erodible sandy or chalky soils kept under permanent pasture.

7.2 AGRICULTURAL CHANGE

In the past three or four decades agricultural practices have changed greatly and more intensive technological agriculture no longer automatically sustains this pleasant rural environment. Farming after World War II has been supremely successful in achieving its paramount objective of increasing production; we are now in the UK 80% self-sufficient in temperate foodstuffs compared with 30% before the war and in Europe most foodstuffs are produced in surplus. However, this huge success has been bought at great cost, not only in terms of the very large amounts of public money poured into the industry, but also as major damaging environmental impacts and loss of farm livelihoods. Modern intensive post-war agriculture, supported by state subsidies and technological developments, has been able to overcome most of the old limiting factors to production. Cheap fertilizer has enabled hungry soils to be brought into arable use; powerful machinery has facilitated woodland clearance and made possible the ploughing of steep slopes and electric pumps have extended drainage schemes. Product subsidies, especially for arable crops, have thus distorted the relationship between cropping patterns and inherent land capability. Land best suited to trees or grass has been ploughed and wetlands important in regulating river flows have been drained. As a result environmental problems such as erosion, flooding and pollution have ensued, habitats destroyed, biodiversity decreased and many of the amenities of the countryside lost. There has also been a substantial drift of farmers and farm workers from the land, related in part to a reorganization of farm structures into larger more economically efficient units. The contribution of agriculture to the economy of rural communities has consequently greatly declined (Shoard, 1980; Body, 1982).

> Farmers and foresters are unconsciously the nation's landscape gardeners . . . even were there no economic, social or strategic reasons for the maintenance of agriculture, the cheapest way, indeed the only way of preserving the countryside in anything like its traditional aspect would still be to farm it.
>
> Scott (1942)

> For the first time in our history there is an increasing divergence between farming on the one hand and nature and landscape conservation on the other, and the farmer, far from being accepted as the guardian of the countryside, is in danger of being regarded as its potential destroyer.
>
> DOE (1978)

Food production is thus no longer so universally accepted as necessarily the best, or only means of maintaining the countryside, nor full-time professional farmers as the sole custodians of the countryside. Local authorities, conservation organizations, game managers and agencies as diverse as the Ministry of Defence and water authorities are also increasingly involved in countryside management which may, or may not, involve the taking of a crop from the land. Yet as agriculture uses so much of the land, farmers must inevitably remain the main agents of countryside management (Green, 1989). Measures have already been introduced to support less intensive, more environmentally-friendly farming in designated Environmentally Sensitive Areas and to modify production control measures such as the Set-Aside and Farm Woodland Schemes in ways which will help them protect and maintain the environment. There seems to be little doubt that environmental objectives will figure largely in the EC Common Agricultural Policy for the foreseeable future (CEC, 1985).

7.3 DILEMMA AND OPPORTUNITY

Current over-capacity within agriculture thus offers an unprecedented opportunity to remedy past environmental damage, restore lost ecosystems and develop sustainable rural land uses which create a rural environment that is not just an accidental by-product of the agricultural industry, but is designed specifically for the social, economic and ecological needs of the late twentieth century. To achieve this we need, above all, some consensus of what is desirable. This is not easy. Integrating agricultural, social and environmental objectives would be difficult, even if there were agreement on what these individual objectives were. There is no such agreement. There are fundamental differences in value systems and perceptions of the underlying issues both within and between all three of these sectoral interests and thus very substantial dilemmas as to the best courses of action. An unprecedented cascade of reports from government agencies, professional bodies, voluntary conservation organizations and academics have set out their own approaches to possible solutions (e.g. Jenkins, 1990; NCC, 1990, Taylor and Dixon, 1990; Marsh *et al.*, 1990; Young, 1991).

Food production might be brought back more into balance with demand by withdrawing any of a number of resources from agriculture, using a variety of possible mechanisms such as quotas, fertilizer taxes, or cuts in support of commodity prices. The nature of the resources withdrawn and the means chosen to withdraw them are major factors influencing the impact of such changes on the countryside and its management. A simple reduction in price support would, for example, probably accelerate the continuing loss of small farms and produce a countryside where production was concentrated into relatively few large, productive farms and much land diverted to other uses or taken out of production altogether. In contrast, control of the use of agrochemicals might favour more a countryside of numerous, small, low input–output farms. These are just two of a variety of possible policy combinations, but are significant in representing two possible extremes in terms of their effects on the environment and farm structures.

Food surpluses can be reduced ultimately by either taking land out of production or by producing less intensively than technology permits. The nature of the future rural environment will depend on which of these strategies, or mix of them, is favoured. In the USA the former strategy has been chosen and, since the 1930s, a series of acreage reduction programmes has converted millions of hectares of cropland to parks, forests and grazing lands. One of the most visited National Parks in the USA, Shenandoah, was created in 1936 by just such abandonment of farmland. It now bears fine stands of second-growth hardwood forest. One only has to visit the East Coast States of the USA, or to read the large American ecological literature on old field successions, to appreciate the rich ecosystems and beautiful landscapes which have been spontaneously developed on old farmland. In Europe, this process has occurred in an unplanned way, even during the high farming since World War II. Between 1965 and 1983 the agricultural area used decreased by 11 million ha (8%) and the extent of forestry increased by 15% in the European Community (CEC, 1988).

Despite the deficiency in timber production in Europe (the Community is approximately 40% self-sufficient and the UK only 10%), this process has not been viewed as positively as across the Atlantic. The European Commission has argued strongly against a European agriculture modelled on that of the USA with 'large reserves of land and few farmers' (CEC, 1988). Yet it then goes on to identify urban fringe, farming heartlands and remote marginal farming areas where, in the first and last cases, the future of farming is perceived as highly problematic. Many agricultural economists favour the economically efficient approach of concentrating production into low cost areas of comparative advantage. Some environmentalists also see great potential benefits in its corollary of diverting large tracts of poor land out of production into other less intensive uses. Such polarization of land use and the loss of numerous, less efficient farm livelihoods is, however, vigorously opposed by the majority of farm and conservation interests. In many parts of Europe (particularly in France), there is already great concern that agricultural abandonment is producing rural depopulation and deserted settlements. There are similar worries in the UK, coupled with the fear that the reduced need to protect land for food production will lead to an irreversible use of good farmland for urban and industrial development. The policy has been opposed both by those wishing to preserve their own pleasant rural living environments and others more preoccupied with the need to retain the capability in the agricultural industry to restore production readily, should for example, global climatic changes eliminate present food surpluses. Reflecting pressures from the former lobby, a recent Department of the Environment circular to planning authorities has, for the first time, introduced guidelines for the protection of the countryside 'for its own sake'.

There is thus a powerful body of opinion that believes the best means of protecting the countryside and maintaining rural communities is to adopt a strategy of returning to more sustainable, lower input–output farming systems. This view is strongly supported by conservation and consumer organizations concerned about a wide range of issues including chemical residues in food; animal welfare in intensive production systems; pollution from nitrates, pesticides and other agrochemicals; and the barren landscapes, devoid of wildlife and recreational opportunity, that intensive production generates. A resistance to change and to the abandonment of hard-won land back to wilderness are also powerful factors underlying this approach.

Despite the collapse of the consensus view of the farmer as the custodian of the countryside, these debates are now beginning to lead to an uneasy coalition between farmers and conservationists, a coalition united against the economically-efficient solutions to over-capacity and pledged to maintain existing farm structures through subsidies to protect farm livelihoods and manage the countryside. This is a strange phenomenon, given the differing views widely held until so recently by farmers and conservationists. Such differing positions, or opinions, arise from differing perceptions of the environment which have become institutionalized into received wisdoms forming the basis of core axioms and overarching principles (even ethics?) on which actions and policies are based. Their development thus bears some examination.

7.4 PERCEPTIONS OF NATURE

Unlike most conservationists, who commonly see nature as fragile and life-sustaining, many people, particularly farmers, still see nature as hostile and resilient; something that has to be fought to keep under control. They worry that one only has to leave a garden or field untended for a few days and brambles, nettles, thistles and all manner of other undesirable plants and animals come leaping out of the hedgerows to reclaim hard-won land back to the wild. This antagonistic frame of mind at war with nature is very clearly reflected in the trade names of herbicides: '*Avenge, Assassin, Impact, Lasso, Vulcan, Musketeer, Gladiator,*

Marksman, Kombat, Topshot, Javelin, Commando, Dictator, Dagger, Stomp, Swipe, Kill, Missile, Fusillade, Clout, Ambush'. This is an ancient, atavistic view of nature, reflecting a warface once vital to our survival. Wilderness literally means wild-deer-ness, or place of untamed wild beasts. Even in the UK the primeval forest would have contained fierce bears, wild cattle and wolves. A mid-seventeenth century dictionary included in its epithets of a forest: *dreadful, gloomy, wild, desert, uncouth, melancholy, unpeopled, beast-haunted* (Thomas, 1983). Control of the wild was important to our very existence, and beauty was seen in the manifestation of this control. *'Every valley shall be filled and every mountain and hill shall be brought low, and the crooked shall be made straight and the rough ways shall be made smooth'* (Isa. 40.4; Luke 3.5). That is not a policy statement setting out the creed of a modern agribusiness, although someone who has seen the big yellow machines grading the face of the North Downs might be forgiven for thinking so. It is the survival imperative of an ancient people. It is no coincidence that early gardens were designed with straight geometric layouts emphasizing man's mastery over nature.

When the countryside began to be brought more widely under cultivation, in the UK largely as a result of the Parliamentary Enclosures between 1750 and 1850, informal landscaped gardens mirroring the wild began to be designed by Kent, Brown and Repton, reflecting a fundamental change in prevailing attitudes towards nature. As late as the 1820s a perceptive, sympathetic observer could still describe the countryside around Hindhead in Surrey as *'the most villainous spot that God ever made'*; with its *'rascally heaths'*, it was *'very bad land and very ugly country'* (Cobbett, 1830). It is now statutorily designated as an Area of Outstanding Natural Beauty. Perceptions changed dramatically when Wordsworth, Clare and others began to extol the virtues of the wild countryside that was being lost through agricultural change:

> . . . Where bramble bushes grew and the daisy gemmed in dew
> And the hills of silken grass like to cushions to the view.
> Where we threw the pismire crumbs when we'd nothing else to do,
> All levelled like a desert by the never-weary plough . . .
>
> . . . Enclosure like a Bonaparte let not a thing remain,
> It levelled every bush and tree and levelled every hill
> And hung the moles for traitors – though the brook is running still
> It runs a naked stream cold and chill.
>
> John Clare, Remembrances (*c*.1832–35)

It was Wordsworth (1810) who first formulated something approaching the concept of a National Park in the Lake District. The beginnings of the conservation movement date from this time; the Commons, Open Spaces and Footpaths Preservation Society was formed in 1865 to protect some of the last commons so that their amenities could still be enjoyed by an increasingly urban public.

That the older view still strongly prevails among farmers and landowners is not surprising. In Western cultures utilitarian attitudes towards nature have always predominated. This has been attributed to Judaeo-Christian teaching based on injunctions in the Bible which can be readily interpreted as giving man complete dominion over the natural world and authority to exploit it for his own ends (White, 1967):

> *And God blessed them, and God said unto them, be fruitful and multiply, and replenish the earth, and subdue it . . . (Gen. 1.28)*

> *And the fear of you and the dread of you shall be upon every beast of the earth, and upon every fowl of the air, upon all that moveth upon the earth, and upon the fishes of the sea; into your hand are they delivered. (Gen. 9.2)*

There is also support in the scriptures for the very opposite view, of stewardship, namely that it is incumbent on man to look after the environment:

> *And the Lord God took the man, and put him into the garden of Eden to dress it and to keep it. (Gen. 2.15)*

This delight in nature and a longing for a time when man lived in harmony with it has always been an undercurrent flowing counter to the prevailing exploitative utilitarian attitudes. From the biblical Song of Solomon to the present day, wild plants and animals have been dominant in the imagery of poetry, painting and music. Sir Julian Huxley in his preface to Rachel Carson's book, *Silent Spring*, which drew attention to the effects of pesticides on the environment, said that after reading it his brother Aldous, had felt that 'we are losing half the subject matter of English poetry'. This romantic tradition is still a countervailing balance against utilitarianism. It is often claimed that cultures which have adhered to ideologies such as Taoism and Buddhism that see man as part of nature and conforming to her laws have had less damaging effects on the environment. Hunter-gathering peoples, such as the Red Indians of North America, commonly had a profound reverence for nature and seem, even today, to live in some harmony with their environment.

Historically it was the Neolithic revolution of agriculture which overthrew this harmony. Nature now became valueless unless controlled to produce for Man; no longer sacred but merely a platform and fuel for his activities. It has been argued that Christianity legitimizes and promotes this belief and the 'fertility mania' of agriculture and can be seen as the 'conceptual apotheosis of the Neolithic revolution' (Oelschlaeger, 1991). It has also been suggested that Western exploitative philosophy derives from an origin as nomadic peoples. The argument is that nomads, who soon move on, have less of a stake in sustainable exploitation of their surroundings and ecological prudence than more settled peoples (Gadgil, 1985). It is, however, a moot point whether the apparent integration of some primitive peoples with their environment is a result of their ideology, or just the lack of the means to be more exploitative. There are protected forests in the vicinity of shrines and monasteries in China, but little evidence elsewhere of a sustainable relationship with the environment of a kind consistent with Buddhist or Taoist philosophies. Wherever populations have developed, or imported, the means to overcome the local environmental constraints to their growth they have generally done so at the expense of their former more integrated relationship with the environment. Are ethics, maxims and the dictums of organized religion merely the rationalizing of these necessities to counter nostalgia for a bygone Arcadia? Or are religious beliefs 'really enabling mechanisms for survival'? (Wilson, 1978).

7.5 STEWARDSHIP AND DOMINION

The old duality between stewardship and dominion is as potent a force today in determining man's relationship with the environment as it has always been in the past. Some substantial mental gymnastics are taking place in the farming community to adjust to their new role as steward and countryside manager. Farmers once scornful of a role as 'park manager' are now becoming reconciled to this function as it becomes clear that it is a publicly acceptable way in which subsidy to some farming sectors can continue. The switch from the view of the countryside as a factory for the production of food and fibre to that of a place of spiritual inspiration and renewal, of freedom and pleasure in wildlife and beautiful scenery is one which some will find difficult to make. Environmental protection is a difficult concept to people like farmers and foresters who appreciate the countryside in a functional way and see change as a desirable, indeed inevitable, result of man's essential mastery over nature. To them there is merit in whatever kind of landscape results, provided gross misuse is avoided and the land remains 'in good heart'. This kind of attitude was summed up in a statement by Jerry Wiggin which he made in 1981 as Parliamentary Secretary to the Ministry of Agriculture, Fisheries and Food:

> . . . there is no such thing as the natural beauty of the countryside. The countryside as we see it today has been made by the efforts of countless generations of landowners and farmers since this land was first inhabited, and the landowners and farmers of today are just as conscious of the beauty of the countryside around them as were their forefathers. It is totally untrue therefore – and I reject utterly the

> accusation – that farmers and landowners are damaging the countryside. Of course they are not. The countryside is a living, ever-changing entity and I know of no greater conservationists than those people who live in it and who eke a living from their work within it.
>
> The countryside is often portrayed as a timeless unchanging environment. It is a false image. The countryside has been moulded by man's activities. As those activities have changed, so has the countryside. Change, not inertia, has been its most abiding characteristic.
>
> Royal Institute of Chartered Surveyors (1986)

In contrast conservationists are more likely to perceive continuity in the countryside and wish to protect it as little changed as is possible:

> Except for town expansion, almost every hedge, wood, heath, farm, etc. on the Ordnance Survey large-scale maps of 1890 is still there on the air photographs of 1940 . . . Much of England in 1945 would have been instantly recognizable by Sir Thomas More, and some areas would have been recognized by the Emperor Claudius.
>
> Rackham (1986)

The preservation of the present farmed, manicured landscape is a very strong motivation in the membership of some conservation organizations. Farmers are now being paid to maintain the cherished flower meadows and walled landscape of the Yorkshire Dales National Park, for example. But this landscape only dates from the Parliamentary Enclosures and replaced a quite different earlier walled landscape. Would the National Park Authority have objected to that change if it had been present when the Parliamentary Enclosures were being planned? I think the answer is almost certainly 'yes'. And why do we object to derelict stone barns in the landscape, but cherish derelict abbeys, or even derelict Cornish tin mines? There would seem to be no reason why change has necessarily to be for the worse. For many of those to whom the countryside is essentially an aesthetic experience, the ideal is often one of 'wilderness' where there has been no human impact whatsoever. Ranged between these two polarities there is a whole series of positions which see both utilitarian and aesthetic benefits as being maximized at some intermediate, harmonious level of exploitation by man. Those who think this way are usually prepared to accept the inevitability of some change in the countryside, but wish to control it in order to maintain those qualities thought to be most desirable. It is this position, reflected in the significant change of name of a key conservation organization from the Council for the *Preservation* of Rural England to the Council for the *Protection* of Rural England, which is beginning to make possible the rapprochment between farming and conservation.

There are, however, also differences of scale as well as quality in the way different people perceive the countryside. People do not only sense merit in quite different attributes of the countryside but perceive the environment at different levels. Some see much more detail than others. Thus the bird watcher becomes very concerned when a single species such as the sparrow-hawk is lost from an area because of pesticides. Other countryside users may not notice, may not even be aware of the existence of such a species. But they would notice the loss of a copse where the sparrow-hawk nests and hedgerows where it hunts. Such changes might not bother the average farmer, but he certainly would be concerned at changes such as soil erosion which they might bring about.

Such considerations apply strongly to the effects of the use of agrochemicals. The farmer, encouraged by the manufacturer, is likely to take the view that such products are innocent of adverse environmental impacts until proven guilty. The environmentalist will usually take the opposite, proactive view. The reactive, product-based approach to legislation has contributed to major environmental catastrophes (e.g. the declines in birds of prey and other species, such as the otter), which were consequent on the introduction of the chlorinated hydrocarbon pesticides in the 1950s. Such experiences are now leading to a more proactive approach based on prudence and the precautionary principle. This seems only common sense when applied to the release of genetically-manipulated organisms, but in other cases the costs of prudence can also be very high. It is estimated for example that the difference in the cost of nitrate reduction measures over the next 10 years in the UK to achieve the EEC Drinking Water Directive Standard of 50 p.p.m. rather than a derogated standard of 80 p.p.m. would be £100 m; money which might achieve much for the environment elsewhere (DOE, 1986).

7.6 FARMING AND CONSERVATION

Quite apart from these differing perceptions of, and positions on, environmental issues between farmers and conservationists, there are very real practical difficulties in marrying farming and conservation. High production ecosystems are almost invariably poor in species, whether reed beds or wheat fields. It has long been known from field trials such as the Park Grass plots at Rothamsted that fertilizer is a very effective way of reducing plant and animal diversity. Unless, therefore, some high value, low output product can be raised, environmentally friendly farming systems are going to be very expensive to support. With tourism as our biggest industry, farm-based holidays and other recreational activities may be an important way of doing this. Like the new role of countryside manager, however, this prospect clashes in many ways with the puritanical virtues of prudence, hard work and the suspicion of leisure and pleasure which characterize many farming communities.

For such reasons it is taken by many as axiomatic that a beautiful landscape can only result from a healthy rural economy, as the recent Department of Environment draft policy guidance note avers (DOE, 1989). This received wisdom of agricultural fundamentalism stems from the influential Scott 1942 report on Land Utilization in Rural Areas quoted earlier in this chapter. It is this view that has led to the support of economically inefficient farming in many marginal areas such as the uplands. This report also asserted that: '*A radical alteration of the types of farming is not probable and no striking change in the pattern of the open countryside is to be expected.*' We know now how agriculture changed and what a successful food production industry has done to landscape beauty, wildlife and access. The early landscape designers, who lived through the effects of the Parliamentary Enclosures and revolutionized garden design away from the tamed to the wild in direct response to them, were under no such illusions. Thus Humphrey Repton: '*the beauty of pleasure-ground, and the profit of a farm, are incompatible . . . I disclaim all idea of making that which is most beautiful also most profitable: a ploughed field and a field of grass are as distinct objects as a flower garden and a potato ground*', and William Gilpin: '*Land which is merely fertile is a barren prospect*' (quoted in Newby, 1979).

Some landscapes and semi-natural habitats need agricultural management so that farming is bound to be the most effective force for managing most of the countryside for amenity as well as for food production. Overcapacity within the industry offers the opportunity to farm less intensively in ways which will help sustain the livelihoods of small farmers, maintain rural communities and protect the environment. But it does not follow that farming is needed everywhere to maintain these values. The other main way to cut back production is to take land out of farming altogether. In many marginal farming areas, such as the hills, where some farms are only maintained by subsidies and successors to the current generation of farmers are scarce, this may become much more widespread. In the UK much of the land surplus will likely prove to be grassland. If ethical objections on grounds of animal welfare do not prevent the introduction of new technologies (such as BST), improvements in animal production might be such that stock numbers needed for food will be so low as to make it difficult to graze existing grasslands at a level sufficient to maintain them. If animal production in Less Favoured Areas is not supported specifically for social and environmental reasons relatively little land would be required in the lowlands to replace all their production. Forest might thus again eventually advance at the expense of grassland. Whether this would be regarded as desirable environmentally is controversial. We are very fond of our open landscapes and the access to big open spaces that they provide.

As in North America, the future of much of our farmland may thus be a return to the forest and fen from which it was originally carved. The Countryside Commission's proposed New Midlands Forest and Community Forest proposals have been its most imaginative and widely welcomed recent initiatives. The farmed margin has always advanced or retreated according to the economic dictates of the times. We may now again see the white, eroding arable of the chalk reverting to down, pasture to woodland and drainage systems being allowed to decline and land flood back to fen and marsh. Such restored environments, managed by a much more multipurpose agriculture and forestry, could help prevent the pollution of aquifers, control erosion, limit flooding and constitute an important recreational resource. Land uses might again be more appropriately related to inherent land capability; the exploitation of areas of comparative advantage leading to a convergence of economic and ecological efficiency. In such a resource-based system of land planning and management, traditional goals of prosperity and equity might come to be superseded by others such as sustainability, diversity, reversibility, even perhaps austerity. Rethinking and remaking policies for agriculture and the countryside at this scale is the great challenge facing agriculture and the conservation movement. To do it we must re-examine old attitudes, reassess objectives, and decide what kinds of industry and countryside are wanted for the future (Sagoff, 1989). Agriculture may be the first major industry to have reached the plateau of diminishing returns to material and technological progress. New objectives will mean new ethics.

REFERENCES

Body, R. (1982) *Agriculture: the Triumph and the Shame*, Temple Smith, London.

Commission of European Communities (1985) *Perspectives for the Common Agricultural Policy*, COM(85)33 final, Brussels, Luxembourg.

Commission of European Communities (1988) *The future of rural society*, COM(88) 501 final, Brussels, Luxembourg.

Cobbett, W. (1830) *Rural Rides*, Penguin (1987), London.

Department of the Environment (1978) *Food production in the countryside*, Topic Paper No.3, The Countryside Review Committee, HMSO, London.

Department of the Environment (1986) *Nitrate in water. A report by the Nitrate Co-ordination Group*, Pollution Paper No.26, HMSO, London.

Department of the Environment (1989) *Countryside and the rural economy*, Draft Planning Policy Guidance Note, PPG 7, HMSO, London.

Gadgil, M. (1985) Cultural evolution of ecological prudence. *Landscape Planning*, **12**, 285–99.

Green, B. (1989) *Countryside Conservation: the Protection and Management of Amenity Ecosystems*, Unwin Hyman, London.

Jenkins, T.N. (1990) *Future Harvests: the Economics of Farming and the Environment; Proposals for Action*, CPRE, WWF, London.

Marsh, J., Green, B., Kearney, B. *et al.* (1990) *A Future for Europe's Farmers and the*

Countryside, Land Use and Food Policy Inter Group, European Parliament, Strasbourg.
Nature Conservancy Council (1990) *Nature Conservation and Agricultural Change*, NCC, Peterborough.
Newby, H. (1979) *Green and Pleasant Land?* Hutchinson, London.
Oelschlaeger, M. (1991) *The Idea of Wilderness*, Yale University Press, New Haven.
Rackham (1986) *The History of the Countryside*, Dent, London.
Royal Institution of Chartered Surveyors (1986) *Managing the Countriyside*, RICS, London.
Sagoff, M. (1988) On teaching a course on ethics, agriculture and the environment. *J. Agr. Ethics*, **1**, 69–84.
Scott, Lord Justice. (1942) *Report of the committee on land utilisation in rural areas*, Cmd 6378, Ministry of Works and Planning, HMSO, London.
Shoard, M. (1980) *The Theft of the Countryside*, Temple Smith, London.
Taylor, J.P. and Dixon, J.B. (1990) *Agriculture and the Environment: Towards Integration*, RSPB, Sandy.
Thomas, K. (1983) *Man and the Natural World*, Allen Lane, London.
White, L. (1967) The historical roots of our ecologic crisis. *Science*, **155**, 1203–7.
Wilson, E. (1978) *On Human Nature*, Harvard University Press, Cambridge, MA.
Wordsworth, W. (1810) *Guide to the Lakes*, Oxford University Press (1977), Oxford.
Young, J. (1991) *Changing Countryside Policies*, RICS, London.

Case study: farm animals

8

R. Harrison

Ruth Harrison is the author of *Animal Machines* (Vincent Stuart, 1964), and since its publication has been an important influence in the monitoring and improvement of conditions for farm animals. She is a member of the UK Farm Animal Welfare Council (FAWC).

8.1 INTRODUCTION

In 1960 I received a letter containing a leaflet that dramatically changed the course of my life. On the front of the leaflet was a picture of a veal calf standing in a crate and emblazoned over it was the question '*Cheap* food yes, but is it *good* food?' (Crusade Against all Cruelty to Animals). This started me on a course of investigation which culminated in my writing *Animal Machines* (Harrison, 1964). The book looked at the ways in which farm animals were kept, fed and slaughtered; at the many drugs and additives used; and at the resulting quality of the food produced. It also looked at the effects on the environment and the problem of world hunger. The conclusion was that it was not *good* food, and that change was urgently needed.

Before going any further, it is worth looking at the development of factory farming. The systems were developed in the years after World War II when governments were pledged to improve on the austerity of the war-time diet. As there was also an acute shortage of labour on the land, the whole thrust was for greater productivity and increased mechanization. Two things made the systems possible: the application to agriculture of antibiotics which enabled farmers to multiply the number

Environmental Dilemmas Ethics and decisions
Edited by R.J. Berry
Published in 1992 by Chapman & Hall, London. ISBN 0 412 39800 1

of animals they could keep in buildings, and the derationing of feedstuffs in 1953.

Specialized buildings, with standard pens, stalls and cages, lend themselves to automation, easing the producer's workload even further, and this became more and more sophisticated over time. The design of systems moved out of the hands of stockmen and into the hands of engineers and technicians, men of great skill and ingenuity, but with little knowledge of animals and in particular of animal behaviour. Human gain was the sole aim of this single-minded pursuit. There was no intention at all of harming the animals.

Thus a large proportion of our food animals, but by no means all of them, lived in small pens or cages, standing on slatted or perforated metal floors, unable to turn round, unable freely to stretch their limbs, their only relief from boredom the passing stockman or the arrival of monotonous mash or pellets at feeding time. To counteract the aggression that can arise from such conditions many of the animals were kept in dim light or even in darkness; and they were mutilated: hens debeaked, piglets' tails docked, and cattle dehorned.

Perhaps it would be as well to point to a clear distinction between 'intensive farming' and what is called 'factory farming'. The intensive farmer uses technology with discrimination; he takes advantage of increasing knowledge and research to achieve greater productivity from his land by better management, manuring, feeding and breeding, *but without significantly changing the pattern of life his animals lead.* The factory farmer aims at a maximum turnover of capital with a minimum of effort. He makes use of all that technology has to offer him without serious question, and *he places the animals in a completely strange situation to which they are not adapted by either domestication or evolution.*

Thirty years ago, the chief Act protecting these animals was the 1911 Protection of Animals Act, which makes it an offence to cause 'unnecessary suffering' to any animal under man's control, an exception being made for slaughter. This definition mainly covered acts of physical brutality but it also included the more obvious mental suffering: to 'infuriate or terrify an animal'. Case law had established that the suffering had to be substantial for a prosecution to succeed.

This was the situation that faced me, and in turn faced the Brambell Committee.

8.2 THE BRAMBELL COMMITTEE

My book was serialised in *The Observer*, and publication was followed by an outcry both in the press and in Parliament, focusing on the plight of the animals. The Ministry of Agriculture took the unprecedented step of calling a press conference to defend the practices under criticism in order to allay public concern. The Ministry's main contentions were that animals would not thrive if they were suffering and that the infliction of boredom did not constitute cruelty. Its Chief Scientific Officer claimed that 'no case had been established for making it an offence merely to deprive animals of light, freedom to exercise or pasture'. He was speaking of the legal position but even so the word 'merely' attached to such deprivations did not impress people, nor did his following comment that neither, in his view, did they cause suffering (Turner, 1964).

Within six weeks of publication the Minister announced that he was setting up a Technical Committee to look into the animals' welfare and to make recommendations where necessary (Brambell, 1965). This committee became known as the Brambell Committee after its Chairman, Professor Rogers Brambell, a distinguished Professor of Zoology. The Committee was composed mainly of agriculturalists but it had the advantage of having an ethologist of international standing, Dr (later Professor) William Thorpe. The Committee visited farms and consulted widely both at home and overseas. Their Report was published at the end of 1965.

The Committee did not share the Chief Scientist's complacency. They pointed out that consideration for animals was an evolving process and that conditions which appeared acceptable today would probably appear unacceptable in the future. Welfare was defined not only in terms of health, but as being 'a wide term that embraces both the physical and mental well-being of the animal. Any attempt to evaluate welfare must take into account the scientific evidence available concerning the feelings of animals that can be derived from their structure and functions and also from their behaviour'. They pointed out that just as the feelings of another individual can only be evaluated by analogy with one's own feelings, so those of an animal must similarly rest on analogy with our own and must be derived from observation of the cries, expressions, reactions, behaviour, health and productivity of the animal.

Proposals were made for regulations designed to prevent existing suffering on farms, and to give effect to the minimum requirement set out in the Report:

> An animal should at least have sufficient freedom of movement to be able without difficulty, to turn round, groom itself, get up, lie down and stretch its limbs.

These became known as the 'Brambell five freedoms' and, in turn, sprang from their fundamental principle that they:

> . . . disapproved of a degree of confinement which necessarily frustrates most of the major activities which make up its natural behaviour.

An appendix to the Report, written by William Thorpe, collated existing information on the behaviour of the main farm species. He stressed that 'all the domestic animals which man farms are species which, in the wild, show a fairly highly organized social life, either in flock, family, clan or herd. This means that their mental and behavioural organization is also potentially on a high level, far higher in fact than the ordinary man imagines'.

Other recommendations for immediate implementation included:

1. There should be sufficient light in animal houses for routine inspection;
2. Manufactured milk substitutes for calves should in no way be deficient in iron, and calves should receive palatable roughage daily;
3. Periodic bleeding of veal calves (to cause anaemia) should be banned;
4. All housed cattle and calves should have a bedded lying area;
5. The mutilations of debeaking of poultry and docking of pigs' tails should be banned;
6. The use of spectacles and blinkers on poultry should be banned;
7. Battery cages should be banned within 2 years.

Professor Thorpe described the anatomy of the bird's beak in his appendix to the report, pointing out that there were nerve endings within a few millimetres of the end of the beak. This was dismissed by some scientists who sided with industry in declaring that cutting off a part of the beak was no more severe an operation than cutting one's finger nails. It was not until the 1980s that Thorpe was vindicated by the work of Gentle and Breward (1981), who have shown that not only does the debeaking operation itself cause pain, but that neuromas form on the site of the amputation and the birds may suffer the same long-term trauma as people with amputated limbs.

Longer term recommendations were:

1. For a new Act giving a fuller definition of suffering, and in particular to cover behavioural distress;
2. The intentions of the Report should not be prejudiced by imports of food produced under unacceptable conditions;
3. For a Standing Committee to carry on where they had left off – this committee, like their own, to contain no person with commercial interests;
4. There should be more behaviour studies carried out on farm animals;
5. Work to be undertaken as a matter of urgency into developing and improving loose housing systems to replace those of extreme restriction;
6. There should be more careful selection and training of stockmen.

In their response to the Brambell Report (1966), Ministers promised four regulations:

1. Control of the docking of pigs' tails;
2. A ban on the reintroduction of the old custom of bleeding veal calves;
3. Sufficient light for routine inspection of pig and poultry houses;
4. A minimum level of iron in veal milk substitutes.

Ministers ignored the scientific evidence produced by the Committee and gave lack of scientific evidence as the reason for not making more of the Brambell recommendations mandatory; they also declared that they were not prepared to make money available to obtain that evidence.

To assess how far either the public or the farmers agreed with the findings of the Brambell Committee, Social Surveys (Gallup Poll) Ltd were commissioned to undertake two surveys, one of the public and one of farmers (1969). In the survey of farmers, nearly 1900 were interviewed from all over the UK. Additional interviews were held with specialist dairy and pig farmers and egg producers to provide adequate samples for regional analysis.

Questions to both the public and the farmers covered, first of all, the Brambell Committee's guiding principle:

1. 'All farm animals should be freely able to turn round in their pens (except when tethered or restrained for veterinary treatment or other temporary purpose).'
   87% of the public agreed with this recommendation and
   65% of the farmers.

2. 'All birds should have enough room to spread their wings.'
   91% of the public agreed with this and
   78% of the farmers.

Other questions put to the farmers showed overwhelming support for the Brambell recommendations. For example, 94% thought that every animal should be given a diet to keep it in full health with no deliberate deficiency.

Further questions were put to both the public and the farmers:

1. Whether grazing animals – cattle and sheep – should have access to the open for part of the day in fair weather?
   80% of the public thought they should, and
   76% of the farmers.
2. Whether all other stock, including poultry, should have access to the open for some part of the day in fair weather?
   79% of the public thought they should, and
   56% of the farmers.

In spite of all this support no action whatsoever was taken on the Brambell Committee's recommendations for immediate action to counter existing suffering on farms. There is no doubt that if Government had been brave enough to make the regulations at the time the face of livestock farming throughout the world would have been vastly different today.

8.3 FARM ANIMAL WELFARE ADVISORY COMMITTEE

In 1967 Ministers set up the Farm Animal Welfare Advisory Committee to give them advice covering the welfare of animals on the farm, but they ignored the Brambell Committee's appeal that nobody with commercial interests should be invited to serve on it.

The Slaughter of Poultry Act came into force in 1967 making it an offence for the first time to slaughter poultry without first stunning them (with exemption of slaughter for Jews and Muslims).

Then in 1968 a new Act was passed dealing specifically with the on-farm welfare of farm animals – the *Agriculture (Miscellaneous Provisions) Act*.

> Clause 1 of the Act gave a new definition of suffering as being 'unnecessary pain or unnecessary distress', allowing exemptions for research.
>
> Clause 2 gave the Minister power to make regulations covering every aspect of the animals' life; each regulation requiring Parliamentary approval.
>
> Clause 3 enabled the Minister to prepare voluntary Codes of Practice for the guidance of stockmen. Although these are not mandatory, non-compliance can be used in a court of law as tending to establish guilt when a case is taken under Clause 1.

The Act allowed for the inspection of farms by Ministry veterinary surgeons, and for the Ministry to give advice on matters relating to livestock.

Theoretically this should have been a great step forward. The Brambell Report setting the scene for action, the new Act providing the framework, and the Codes providing the 'teeth'.

However the first Codes of Practice (drawn up by the Farm Animal Welfare Advisory Committee), virtually sanctified existing practice. Their main departures from Brambell were:

1. Calves could be fed entirely on a milk substitute as long as it was 'complete in all known nutrients'. In practice this led to a continuance of the white veal trade;
2. Cattle could be stalled and tethered for long periods;
3. Entirely slatted floors could continue to be used for cattle and pigs;
4. Acceptance of sow stalls and cubicles and tethering of sows;
5. Acceptance of battery cages;
6. Acceptance in poultry of beak-trimming, spectacles and dubbing.

A new definition of suffering, combined with clearly defined Codes, could have proved a powerful deterrent, even though the codes were voluntary, and non-compliance was not in itself an offence. But just as non-compliance could tend to establish guilt when a case is taken for causing suffering, so compliance could be used in defence to help prove innocence. The weak Codes, therefore, robbed the Act of the very teeth they were supposed to give it. This was clear years later when cases were taken to court.

The independent members of the Brambell Committee felt so strongly that the first Codes would do nothing to protect the animals against suffering that they wrote a letter of complaint to *The Times* (1969). Their letter was supported by one from Sir Julian Huxley and nine fellow scientists familiar with the behaviour of animals (1969), who said that it seemed to them that 'behavioural distress had been completely ignored. Yet it is the frustration of activities natural to the animal which may well be the worst form of cruelty'. One of them (J.S.H.), the letter continued,

> would like to confirm this from his experience when in charge of the London Zoo, whose limited extent made a certain degree of crowding inevitable. Animals might be in perfect physical health, and capable of reproduction, but if in solitary confinement, or in too close quarters, without opportunities for play and exercise, would often exhibit signs of dejection or distress. This was not the case at Whipsnade, where there was more space, and the animals enjoyed greater freedom.

Thorpe (1969), dismayed that the lack of scientific evidence was being used to defend the weakness of the Codes, wrote an article in *Nature* setting out the evidence supporting the Brambell minimum standards. In it he said:

> It will be noted that freedom to turn round and stretch concerns in each case restriction of the pattern of locomotion. There is a great deal of information on this topic in the extensive literature concerning experimental neurosis in mammals. There is ample evidence that confinement, just because it restricts what is both physiologically and ethologically one of the most basic and all-pervading activities of animals, namely the locomotory responses, is most powerful in its effects; hence the Brambell Committee's stand against permanent tethering or close confinement of domesticated animals. Moreover, it has long been known that disturbed behaviour caused either by noxious or ambivalent stimuli is greatly enhanced if the stimuli are applied when the animal is in closely confined conditions.

Still the evidence was ignored, although an encouraging outcome of the Brambell Report was the setting up at the end of 1966 of the Society for Veterinary Ethology, whose members meet twice a year to discuss new research findings. It now has an international standing and is a consultant to the Council of Europe.

In his inaugural address to the Society, Professor Brambell again emphasized the importance of behavioural studies on farm animals, which he believed were 'not only important for the welfare of the animal, but also . . . for the welfare of animal husbandry . . . happier and more contented animals are ones that are better doing, but well doing in the true sense is something which is more comprehensive than simple growth, or egg, or milk production, it relates to all aspects of the animal's well-being'.

8.4 FARM ANIMAL WELFARE COUNCIL

In 1979 a change in government in the UK brought a change to the Ministers' advisory committee. The Farm Animal Welfare Council took over from the old committee. Its membership was increased and its terms of reference widened to include the welfare of livestock not only on the farm, but in transit, at markets and at slaughter. Furthermore, it became an autonomous committee with direct access to the public. It later won from the Minister the right to take recommendations direct to the European Commission. A formidable opportunity to improve welfare! On its appointment it issued a press notice saying that it was going to revise the Codes of Practice and wished them to provide farm animals with:

1. Freedom from thirst, hunger or malnutrition;
2. Appropriate comfort and shelter;
3. Prevention, or rapid diagnosis and treatment, of injury and disease;
4. Freedom to display most normal patterns of behaviour;
5. Freedom from fear.

These became known as the 'FAWC five freedoms'. The Council also made the important point that 'animal welfare raises certain points of ethics which are themselves beyond scientific investigation. (It) will therefore especially wish to encourage alternative sysems of livestock husbandry which are ethically acceptable to the concerned public, can be shown to improve the welfare of the livestock in question' – and it then added a curiously self-defeating rider: 'and be economically competitive with existing systems of intensive production', to which it has adhered ever since. Yet surely throughout history, laws have had to be made actually to prevent people using the cheapest ways of doing things.

The revised Codes, issued in 1983, recognized that systems could not be changed overnight and therefore included them in recommended husbandry practice, whilst emphasizing that change was needed. The revised Pigs Code, in particular, had a profound influence on pig farmers.

It pointed out that keeping pigs on slatted or perforated floors can cause severe problems such as lameness or damage to the feet, and it continued:

> Given the opportunity, the pig eats fibrous material, also roots about and makes a nest and uses a separate dunging area. Bedding, and especially straw, contributes towards the needs of the pigs for thermal and physical comfort and satisfies some of its behavioural requirements.

The Code 'strongly recommended the use of straw-based systems'. Later in the Code the keeping of dry sows and gilts in stalls and tether systems was equally strongly criticized for causing welfare problems. Pig farmers began to see the writing on the wall and a steady move started into straw-based systems and away from stalls and tethers.

During the 1980–81 session of the UK Parliament, the Agriculture Committee of the House of Commons decided to do an update on Brambell, confining their studies to pigs, poultry and veal calf production. The Committee published their report in 1981. Like the Brambell Committee, members undertook visits to farms and to research institutes at home and overseas, and took oral and written evidence, but in addition they commissioned a review of the economic consequences of switching to alternative systems. Like the Brambell Committee they came down against battery cages, veal crates, and sow stalls and tethers – all of which 'they viewed with distaste'. The Committee went further than Brambell in their statement of principle:

> We consider that society has a duty to decide upon certain standards in respect of its treatment of animals. We maintain that this can be done without relying on philosophical arguments about animal rights or the nature of animal suffering . . . in our view there are certain decisions of principle that have to be made about the acceptability of practices connected with animal husbandry and about environmental conditions in which animals are kept. Although due regard must be given to the scientific evidence about the various aspects of 'well-being', these standards have to be decided upon in the absence of 'proof' about whether an animal 'suffers' and what its 'natural' behaviour might be.

On the question of economic cost, the Committee took a more radical stand than the Farm Animal Welfare Council:

> We cannot accept that the duty (to see that undue suffering is not caused to animals) should be set aside in order that food may be produced more cheaply. Where unacceptable suffering can be eliminated only at extra cost, that cost should be borne or the product foregone.

Among their more general recommendations were that:

1. Research into behavioural and ethological problems should be allocated the necessary funds.
2. Taxation policy should avoid encouraging undesirable methods of husbandry. Grants and other financial incentives should be used positively to encourage better methods.
3. Agriculture Departments and the Agricultural Research Council . . . should consult FAWC at all stages when research programmes are being formulated or altered, and in drawing up those programmes they should regard animal welfare as of at least equal importance to efficiency of production.
4. Ministers and official thinking should give more weight to animal welfare than seems to have been the case hitherto.

In 1986 the first Chair in Animal Welfare was endowed by the British Veterinary Association's Animal Welfare Foundation and set up at the Department of Clinical Veterinary Medicine, at the University of Cambridge, with Professor Donald Broom as its first holder.

8.5 THE EUROPEAN DIMENSION

In 1972 the UK became a member of the European Economic Community and thereafter found it difficult to act alone on any subject, indeed forthcoming membership was used as an excuse for inaction for 2 or 3 years before that accession.

There has been much interaction between countries since then and much has happened on the international front.

The two most important international organizations covering the welfare of farm animals are the Council of Europe and the EC. The Council of Europe (to which 25 States now belong) was set up in 1949 'to work for greater European unity, to improve conditions of life and develop human values in Europe, and to uphold the principles of parliamentary democracy, the rule of law and human rights. Any European state can become a member provided it accepts these principles'.

The EC, now with 12 Member States, was set up in 1957. The Treaty of Rome lays down its policies. Article 2 states:

> The Community shall have as its task, by establishing a common market and progressively approximating the economic policies of the member states, to promote throughout the community a harmonizing development of economic activity, a continuous and balanced expansion, an increase in stability, an accelerated raising of the

> standard of living and closer relations between the states belonging to it.

Human values are only allowed a very brief airing.

In general terms, therefore, it is the Council of Europe which is concerned with the quality of life, and the EC with distortion of trade.

Council of Europe conventions are open for signature and ratification; signature implies general agreement with the principles of the convention, ratification is binding. The Explanatory Text to the convention for farm animals spells out: 'the parties . . . shall be under an obligation to implement the recommendations, either through legislation or regulations or through administrative practice'.

In the EC, Recommendations have no legal force, a Regulation is binding in its entirety on all member states, and a Directive is binding as to the object to be achieved, but it is left to each member state to make its own national legislation to achieve it. Member states not complying with EC Regulations or Directives may be taken to the European Court.

In the early 1960s the World Federation for the Protection of Animals (now part of the World Society for the Protection of Animals), presented the Council of Europe with a seven point plan for conventions to protect animals. Five of these conventions were completed before the Council ran out of money: those covering the welfare of animals in international transport, on the farm, at slaughter, in research and as pets. The Conventions set out the general principles, which a Standing Committee of ratifying states then elaborates with more detailed recommendations.

The Council of Europe's *European Convention for the Protection of Animals kept for Farming Purposes* was open for signature and ratification in 1976 and came into force in 1978. The first of the general principles it laid down is resoundingly positive:

> Animals shall be housed and provided with food, water and care in a manner which . . . is appropriate to their physiological and ethological needs . . .

This was influenced by the Brambell recommendations which were taken much more seriously overseas than they were in the UK. It was also influenced by the 1972 Animal Protection Act of the former West Germany (which also drew inspiration from Brambell), which lays down that the Act 'shall serve to protect the life and well-being of the animal', and 'well-being' is defined as 'based on the normal and vital functions developing and proceeding undisturbed in a manner which is typical for the species and does justice to the animal's behaviour'. The Act also stipulates that 'the person responsible for the animal shall provide accommodation which takes account of its natural behaviour'.

In 1979 the UK ratified the Council of Europe Convention and the

Minister of Agriculture gave Parliament the assurance, 'Certainly it will be our intention to comply with that Convention in the future'. But still little changed.

The EC sat as an Observer on the Council of Europe Standing Committee until each of its member states had ratified the Convention, when it, too, could ratify and thus become a full Member of the Committee. The Committee has now finalized detailed recommendations covering laying hens, pigs and cattle (excluding calves), and fur-bearing animals. It is now working on those for sheep and goats and will then probably tackle all species of table poultry. It is of benefit to the EC to attend these meetings, for it is here that all the detailed discussions and disagreements take place. It is then easier for them to draw on matters which are generally agreed to form the bases of EC Regulations and Directives.

This sets the broad picture. I have mentioned the impact that the 1983 Pigs Code had on the industry, but there are four other welfare watersheds which deserve a special mention.

8.6 FOUR WATERSHEDS

One which passed almost unnoticed in the UK at the time was a prosecution in 1976, the *National Society Against Factory Farming* v. *Quantock Veal*. It was pointed out in Court that the calves were taken up to slaughter weight in pens 22 inches wide by 5 feet long, standing on wooden slats without bedding, that they could not turn round, could not stretch their legs freely when lying down, had difficulty in grooming themselves, were fed solely on a milk substitute which was deficient in iron, and that they were denied the roughage essential to young ruminants. It was further stated that the calves were kept in darkness apart from twice a day feeding. The producers were not complying with the 1971 Cattle Code which stipulated that the width of the pen should be equal to the height of the animal at the shoulder, and that the level of lighting should be such that all the calves could be clearly seen. The NSAFF did not win the case; but nor did they lose it entirely. The Magistrates thanked the Society for taking the case, and asked them to keep an eye on the system as it was obviously one that could cause suffering. The point was taken by the defendants and the firm switched over to straw yard veal production, starting a continuing trend.

A second watershed came when the West German animal protection societies, impatient for action at farm level, prosecuted a series of cases in 1979 against battery farmers under the 1972 Act. The courts agreed that existing commercial cages were not in accordance with the Act in that the birds were deprived of their inherent behaviour of scratching, stretching, wing flapping and preening. But the battery farmers were exonerated on the basis that as most poultry farmers in Germany kept their hens in battery cages these particular farmers could not possibly have known that it was against the law. So, in practice, little changed in Germany, except that the law was amended to recognize battery cages.

The German Minister of Agriculture, aware that the pressure he was under would not diminish, took his concerns to the EC, to press for a Community decision on the future of battery cages. Only cages for laying hens were considered. The Commission started a series of protracted consultations on whether cages should be phased out, the viability of alternatives to the cage, and on the economic consequences of change. Enormous pressure from industry was brought on the governments of each Member State. The success of all the research on alternative systems was put in doubt, and finally, in 1986, EC Directive 86/13 surfaced (1987) setting out standards for hens in battery cages throughout the Community with no suggestion of an ultimate ban on cages. A space allowance of 450 cm^2 per bird was set; to come into force in 1988 in new cages and in 1995 in existing cages.

The third watershed was again the result of concerted action by animal protection societies. This time the Schweitzer Tierschutz (the Swiss Federation for the Protection of Animals) achieved a referendum on battery cages which resulted in a 10-year phasing out of systems for laying hens which did not include nest boxes and perches – in effect all existing commercial battery cages. The phase out began in 1981 and set the scene for renewed pressure all over Europe. Sweden began a similar 10-year phase out in 1989.

The fourth watershed was in 1986 when the Agriculture Committee of the European Parliament, led by Richard Simmonds, also undertook an investigation into pig, veal and egg production, to which they added the transport of animals. They invited experts on each of these subjects to present evidence and published their report in 1987. They came to similar conclusions as the Brambell Committee and the UK Select Committee and asked the Commission to take urgent action. They passed resolutions calling for:

1. Group housing of veal calves and for their diet to contain roughage and adequate amounts of iron;
2. Battery cages for laying hens to be phased out over a 10-year period;
3. Discontinuation of stalls and tethers and the provision of straw for sows, minimum standards for farrowing crates and a minimum 3-week weaning age;
4. More stringent control and enforcement of laws covering the transport of animals, with incentives to persuade the trade to switch over to a carcase instead of a live animal trade.

This has resulted in EC Directives covering transport, the welfare of pigs and the welfare of calves, to add to that on battery bens mentioned above. The draft regulations have to be agreed between 12 member states with very different problems, and concensus obviously involves lower standards than one would have wished, but nevertheless mean a raising of standards in most areas, and consequently a general raising of standards throughout the Community. There is nothing, other than the dreaded 'distortion of trade' to prevent any member state from keeping to higher standards.

All this is a great deal of action on paper, but how have things actually changed for animals on the farm? Let us take the 1965 Brambell recommendations as our yardstick.

First, their 'five freedoms'. How far have we given all our farm animals sufficient freedom of movement to be able without difficulty to turn round, groom themselves, get up, lie down and stretch their limbs – the Committee's absolute minimum requirements?

8.6.1 CATTLE

Throughout Europe cattle can still be tethered throughout the winter or even throughout the year, and some badly designed cow cubicles still do not permit the animals to stand up or lie down normally. The vast majority of veal calves are still crated. It is salutary to note that in the UK it was only in January 1990 that it became mandatory to provide conditions permitting calves to turn round (SI 2021). The promising recommendations of the draft EC Directive on Calves have not been realized in the final weakened version (91/629/EEC). Although the Directive will permit group housing which can allow the calves the 'five freedoms', it continues to allow producers to keep veal calves in crates which, although a little wider than those used at present, will still only permit the calves to turn round during the first few weeks of their lives and then only where they are not tethered. A draft Regulation on adult cattle has yet to be circulated.

8.6.2 PIGS

Throughout Europe sows and gilts are still confined in stalls and/or tethered in cubicles in which they cannot turn round and cannot get up or lie down normally. In some European countries fattening pigs are also kept in similar conditions. In the UK it will only be mandatory for all sows and gilts to be able to turn round in 1999 (SI 1447). The EC Directive on Pigs (91/630/EEC), like that on calves, was drastically weakened during later discussions and is now disastrous for the stalled and tethered sows of Europe. New constructions and conversions of installations in which sows and gilts are tethered will be permitted until 1996 and the use of tethers for sows and gilts will only be forbidden in 2005. The use of sows stalls continues unchecked, however, and will only be reviewed in 1997.

8.6.3 POULTRY

Throughout Europe laying hens may still be kept in battery cages in which they cannot freely stretch their limbs, and can only turn round and preen themselves with difficulty. Council of Europe recommendations state that birds in battery cages shall be able to stand normally and turn round without difficulty, but the EC Regulation does not live up to this and lays down a stocking rate of 450 cm^2 per bird only to be applied in existing cages in 1995, and restriction of movement is as mentioned above. Discussions have begun again (in 1991) towards revising the Directive, and once more centres on whether to ban cages or only to revise stocking densities. The latter seems the most likely outcome, but this time it is hoped to include an Appendix covering alternative systems for laying hens. It is probably useless to hope that this appendix will aim at standards far-reaching enough to be in keeping with the needs of the hen – so that the hens will not be penalized by a series of short-term inadequate standards, or the producers by having to make constant changes. The European Commission appears unable to give producers the stability of knowing what its ultimate goals are likely to be, even with long phase-in periods.

Of the four regulations promised by Ministers in 1966 in response to the Brambell Report, only three have been made:

1. To control the docking of pigs' tails, in 1974 (SI 798);
2. For the provision of light for routine inspection (but note, not at other times) which came into force in 1991 (SI 1445);
3. A minimum level or iron in veal milk substitutes. This was set out in two different regulations: one which specifically mentions iron but only for singly penned calves (SI 2021), and the other that says that all animals should have a suitable diet (SI 1445). Unless both are challenged I doubt if either will prove effective in protecting veal calves. The EC Calves Directive lays down that sufficient iron shall be included 'to provide a positive state of health', which is open to interpretation.

The practice of bleeding calves, which disappeared when it was found that anaemia (and consequent 'white' meat) could as easily be achieved through deprivation of iron, has never been re-introduced, and the promised ban has therefore never been implemented.

The recommendation providing for palatable roughage for veal calves, has been partially carried out in the UK Calf Regulations (SI 2021) which call for sufficient digestible fibre, and the later regulations (SI 1445), which call for a 'suitable diet' – a beautifully ambiguous phrase! The EEC Directive lays down that 'in order to ensure a positive state of health and well-being as well as a healthy growth rate and to meet their behavioural needs, the calves' food must contain sufficient iron and a minimum of dried feed containing a digestible fibre . . . however, the requirement for a minimum quantity of dried feed . . . *does not apply to the production of calves for white meat*'. The same exception is given in the provision of water: 'All calves over 2 weeks of age must have access to a sufficient quantity of fresh water *or be able to satisfy their fluid intake needs by drinking other liquids*'. So there is little advance for veal calves, the very animals for whom these regulations were brought forward.

In general it is still admissible to keep any animal on unbedded perforated or slatted floors in the UK. The only exceptions are that bedding must be provided for sick animals, for lactating dairy cows and for cows in calving boxes (SI 1445). The EC Calves Regulation stipulates bedding only for calves under 2 weeks of age.

The only other short-term Brambell recommendation was for a ban on debeaking and the use of spectacles/blinkers on poultry. The use of blinkers involving penetration or other mutilation of the nasal septum was banned in 1982 (SI 1884), but it is a disgrace that debeaking (or 'beak-trimming' as it is euphemistically called) is still widely used, with great resistance to any ban being introduced. Even the latest recommendations to come from FAWC only recommends a ban on 'routine, non-therapeutic beak-trimming in 1996' (FAWC, 1991) and in terms which make it difficult to see any actual change in practice. This weak recommendation is opposed by a minority of its members insisting on a total ban on this mutilation (FAWC, 1991). The EC has not yet pronounced on the issue.

All these were recommendations for immediate action!

Now for the Committee's longer term recommendations. I have mentioned the 1968 Act which replaced the old definition of 'unnecessary suffering', by the new definition 'unnecessary pain or unnecessary distress', the word 'distress' being intended to cover both physical and mental suffering, (mental including behavioural suffering). In practice the Courts have tended to interpret the new definition in much the same way as the old, i.e. to accept only suffering resulting in clinical symptoms which they can see – any other form being impossible to quantify. Otherwise the Act could be really effective.

The Brambell Committee had no means of knowing that their aim of not allowing their intentions to be undermined by imports of food produced under unacceptable conditions, would be made impossible by our membership of the EC and its ban on any interference with the free movement of goods (including live animals) within its boundaries. The EC, in turn, is in theory limited in its ability to exclude goods from third countries produced by means they do not like by the GATT agreements. On the international front things do not get any easier.

The formation of the Society for Veterinary Ethology (now the International Society for Applied Ethology) did indeed lead to a burgeoning interest in behaviour studies on farm animals as Professor Brambell had hoped, but far too little work has been done on alternative systems, and indeed it has been estimated that only 3% of UK agricultural research money is spent on projects directly related to welfare (Harrison, 1987), whereas the Agriculture Committee called for expenditure to at least equal that spent on production problems.

Far too little effort over the years has been put into selecting and improving the training of stockmen, as Brambell recommended. Indeed standards have fallen drastically, and people appear to have been chosen for their ability to handle mechanized and electrical equipment rather than their empathy with the animals. This has created a particularly difficult problem with the move towards alternative systems which necessarily need a very high standard of stockmanship.

On the international scene matters get more complex by the day as more and more states join the Council of Europe and apply to join the EC. This makes discussions far more complicated as very different problems are brought into the arena, and of course makes consensus much more difficult. It is a great challenge for the two international organizations, but one which they are well able to face; progress is now inevitable even though it may be very slow.

8.7 ANIMAL WELFARE IN CONTEXT

I have only dealt with a small part of a very large subject, the welfare of farm animals, and this, in turn, is only one area of the great many areas of concern covering man's treatment of animals.

However, welfare cannot be considered in a vacuum. Factory farming has led to pollution and other serious environmental problems; it has led to depopulation of the countryside with many side-effects; it has led to health problems both in animals and man; it has also led to problems for farmers caused by the big monopolies; and it has led to rampant overproduction. What started out perhaps as an altruistic exercise has now become a serious anachronism. In theory we now have time to pause and consider where we are going, but in practice governments are unable to look beyond the next election and appear to be ruled more by disaster containment than by positive planning. They have three big problems to face in agriculture today: to bring supply and demand into better balance, to create an agriculture that can co-exist harmoniously with the environment, and to restrict livestock systems to only those which offer the animal a life of comparative fulfilment. These problems are so interrelated that they could be tackled as one if governments had the courage and the wisdom. But it needs time and the will to resist yet further misdirected technology. Can they really plan for the future or will we just stumble on, the halt leading the blind?

REFERENCES

Brambell, F.W.R. (1965) (Chairman) *Report of the technical committee of enquiry into the welfare of animals kept under intensive livestock husbandry systems*, CMND 28, HMSO, London.

EC (1991) Directive laying down minimum standards for the protection of calves (91/629/EEC).

EC (1991) Directive laying down minimum standards for the protection of pigs (91/630/EEC).

Farm Animal Welfare Council (1991) Report on the welfare of laying hens in colony systems.

Farm Animal Welfare Council (1991) Minority report.

Gentle, M.J. and Breward, J. (1981) The anatomy of the beak, in *Proceedings of the First European Symposium on Poultry Welfare*.

Harrison, R. (1964) *Animal Machines*, Vincent Stuart, London.

Harrison, R. (1987) *Farm animal welfare, what, if any, progress?* UFAW, Potters Bar.

Huxley, J. *et al.* (1969) Lettr to *The Times*, June.

MAFF (1974) The docking of pigs (use of anaesthetics) Order 1974 (SI 1974 798).

MAFF (1982) The welfare of livestock (prohibited operations) regulations (SI 1982 1884).

MAFF (1987) The welfare of battery hens regulations (SI 1987 2020).

MAFF (1987) The welfare of calves regulations (SI 1987 2021).

MAFF (1990) The welfare of livestock regulations (SI 1990 1445).

MAFF (1991) The welfare of pigs regulations (SI 1991 1477).

Social Surveys (Gallup Polls Ltd) (1969) Opinion polls (2) for the Co-ordinating Committee on Factory Farming.

Thorpe, W.H. (1969) Welfare of domestic animals. *Nature*, **224**, 18–20.

Turner, E.S. (1964) *All Heaven in a Rage*, Michael Joseph, London.

Case study: lowland wetland conservation

9

Brian Moss

Brian Moss is Professor of Botany at the University of Liverpool. He was previously at the University of East Anglia, where he was responsible for an important series of experiments on the regeneration of the Norfolk Broads.

9.1 INTRODUCTION

In the first century AD a raised bog near what is now Wilmslow in Cheshire was still actively growing and continued to grow for some centuries. Pollen and other reconstructions suggest a surface of shallow pools and *Sphagnum* hummocks surrounded by the still extensive wild wood of the Cheshire Plain. Unknown events brought a sacrificial victim to this place sometime during the Celtic period. He was ritually murdered by being hit on the head while kneeling in position, by garrotting and by throat cutting. He was then left submerged in one of the acid pools, and the tanning properties of the peat water preserved many of his body features until its discovery in 1984 (Stead *et al.*, 1986). The remains of two further bodies of similar age have been found in Lindow Moss. It is a site of some significance, potentially contributing much to our knowledge of ourselves.

Yet if you visit the bog now, you will not find an ecosystem conserved for biological interest and historic significance but a wasteland of drying peat piles as machines tear out the accumulated layers for bulb fibre. And as the *Sphagnum* peat is used up, the machines will remove the underlying

Environmental Dilemmas Ethics and decisions
Edited by R.J. Berry
Published in 1992 by Chapman & Hall, London. ISBN 0 412 39800 1

fen peat to be mixed with battery chicken droppings for the growing of mushrooms. A part of Lindow Moss is technically a nature reserve of the Cheshire Wildlife Trust. A fence teeters on the vertical precipice where the peat has been cut to the edge and what was a vital bog community is now just a drained birch scrub. Lindow Man is preserved in a controlled environment cabinet in the British Museum; it is right to look after his body but the destruction of the environment which gives his body meaning casts doubt on the nature of some of our modern values.

On the outskirts of Port Moresby in Papua New Guinea, a group of people called the Motu maintain a long tradition by living in houses on stilts in the intertidal zone. They have adopted many of the ways and means of urban society but they still collect shellfish at low tide as a significant source of their food. A few kilometres out to sea, surrounded by coral, is a rocky island, Motupore, skirted by the sand flats built up by tide, seagrass and mangrove. At the landward edge of the flats to the north of the island is a large midden recording the activities of the Motu who lived on the island in former days. The stratigraphy of the midden is fascinating for it records the changing domestic life of a group of people little influenced by the Western world until very recently. The deeper layers, perhaps two centuries old, contain the bones of manatees and crocodiles and the shells of large species of molluscs, accompanied by the remains of simple pottery. The more recent layers lack the bones of large animals or the shells of large molluscs but contain fragments of decorated pottery. It seems that a strategy used by other predators had applied. Large prey were selected, for they gave the best return for the energy invested in hunting or collecting them. But the consequence was a scarcity of such prey and a turning to smaller species; the general or seasonal shortage of food as the population increased may have led to a need to produce marketable pots, exportable in return for food from elsewhere, and hence the change in pottery styles from the plain to the decorated.

The midden on Motupore Island lies next to a large flat area which is a burial ground for the Motu. It is extensive and must contain a wealth of material which archaeologists, appetites fuelled by the fascination of the midden, would be keen to explore. But investigation would involve disturbance to the bones of ancestors whom the present-day Motu feel should not be troubled. The site remains uninvestigated, not through legal protection but through the value system of the local people.

These two contrasting examples contain the central dilemma about the preservation and conservation of wetlands: the conflict between immediate need and wise exploitation suggested by the local extinction of crocodiles and manatees by the Motu and the peat mining at Lindow Moss. And they illustrate the human contexts of this dilemma in the different values collectively held by the English and the Motu in the destruction of Lindow Moss and the reverence for Motupore Island. This discussion thus has two foci: local wise use and global values.

9.2 THE NATURE OF WETLANDS

Wetlands include any part of the landscape which regularly has a water table at or above the soil or sediment surface at some time of the year and in which the major primary producers are aquatic plants or non-planktonic algae. They may thus include tundras, muskeg, swamps, bogs, fens, marshes, carrs, seaweed-dominated shores and mudflats and are initially best characterized by the nature of their vegetation. This has been attempted floristically in the phytosociological tradition for particular groups of wetlands, such as fens and salt marshes (Wheeler, 1980) but a physiognomic scheme is perhaps more useful for the present purposes.

Wetland types vary continuously and mutually along several main axes: latitude and plant growth form, mineral nutrient supply, salinity, and hydrology. Superimposed on all of these is the axis of human usage with pristine wetlands at one end and those obliterated by drainage or infill at the other.

The most extensive wetlands are those of the tundra and the muskeg in North America and Eurasia. Permafrost in the tundra prevents free drainage, and the accumulation of water in summer over large flat areas dominated by sedges, grasses and lichens produces meadows and shallow pools, the whole being underlain by a thin peat, reflecting low productivity in the community as a whole. To the south, the muskeg with a sedge or *Sphagnum*-dominated ground flora and a tree cover of gymnosperms and waterlogging-tolerant angiosperm trees owes its existence as a wetland to the excess of precipitation over evaporation at high latitudes. It may bear greater thicknesses of peat in extensive boglands blanketing the terrain. Overall, tundra and muskeg occupy a significant percentage of the land surface, all of it in relatively thinly populated regions.

Further south in the northern hemisphere the deciduous forest biome contains a variety of wetlands, although none of them in such a large and continuous block as the tundra and muskeg. As 'climax' communities they are generally tree dominated by for example, willow (*Salix* spp.) and alder (*Alnus* spp.), or cypress (*Taxodium distichum*) in the warmer temperate USA although natural phenomena such as wind in maritime regions and fire in the warm temperate zone may prevent tree growth to give bogs dominated by low-growing shrubs such as the heaths

(Ericaceae), or sedges such as the saw sedge (*Cladium jamaicense*) of the Florida Everglades (Douglas, 1947).

In the tropics, despite the high evaporation to precipitation ratios, wetlands are abundant although again patchy in distribution, being associated with the floodplain stages of the rivers (Wellcome, 1979). The major changes in water level throughout the year give complex flood plain systems that are generally not wooded but which comprise sedge- and herb-dominated, often floating, permanent swamps like the Nile Sudd and seasonally flooded grasslands at the edges of the valleys.

A second axis of variation, concerning mineral supply, particularly preoccupies observers in countries with a varied geology in the deciduous forest biome. The water supply to a wetland comprises a mixture from two main sources, direct precipitation and precipitation which has passed over rocks and through soils, thereby becoming changed chemically. Direct precipitation (ombrogenous or ombrotrophic supply, from Gk. *ombros* = rain) is relatively low in pH and dissolved substances whereas the ground source (soligenous or minerotrophic supply, from Latin *solum* = soil) varies a great deal dependent on local geology but is generally of higher pH and mineral content. Some soligenous supplies in mountain regions or those of sands and gravels may, however, have a similar composition to rain and those that have filtered through the anciently weathered lateritic soils of the wet tropics may be similarly low in minerals and nutrients.

In the tundra and muskeg, the water supply is predominantly ombrogenous because the permafrost and low evaporation rates essentially lead to a layering of direct precipitation on any underlying ground water and an isolation by the peat of underlying mineral-rich rock debris. It is possible therefore in western Ireland (Bellamy, 1986) to have an acid bog vegetation dominated by ombrotrophic water overlying otherwise readily weatherable limestone rocks which in other circumstances would give extremely mineral-rich water.

In the tropics, peat may form in permanent basins but is oxidized almost as rapidly and so builds up slowly and is readily destroyed by fire. There is thus little isolation from the underlying mineral debris and the main water supply to the wetlands is riverine and derived from the catchment. Some wetlands in the inland tropics may even be highly saline owing to very large evaporation rates.

In the intermediate, temperate zone, however, the pattern is more complex. There is ombrogenous dominance in the cooler or wetter parts leading to acid bog vegetation in basins set among poorly weathered igneous and metamorphic rocks. Drier parts, and those with sedimentary rocks, however, show greater minerotrophic dominance with base-rich fens developing instead of acid bogs. Thus in the UK, bog vegetation is characteristic of wet sites in the west and north owing to the combination of igneous/metamorphic geology and the prevailing north-westerly winds from the Atlantic and fen vegetation characterizes the south and east with their soft rock geology and drier weather (Godwin, 1975). A similar change occurs in the northern USA with bogs found on the maritime seaboards of New England and the Pacific North West and willow-dominated fens in the interior mid-west states like Wisconsin and Illinois south of the muskeg influence.

The influence of salinity, the third axis, is particularly expressed in the distinctiveness of its several communities. There are those seaweed and microalgal-dominated communities of rocky intertidal shores, the algal-dominated mudflats at lower tidal levels on low-energy depositional coasts and the saltmarshes, seagrass meadows and mangrove swamps of the upper parts of these shores. These habitats, together with coral reefs, where possible classification as a wetland illustrates the difficulty of drawing any boundaries in biological systems, occupy all of the world's shorelines that have not been developed for human activities. Because of the enormous total length of shorelines, the significance of the coastal zone as the most productive part of the ocean, and the accumulation of human population towards the coast for historic reasons of colonization and communication, coastal wetlands are extremely important.

The fourth major natural axis is that of hydrology with the poles of the axis being permanent inundation at one end and a seasonal complete drying out at the other. The northern bogs lie at the former end, the many floodplain systems of the tropics at the other. Floodplains comprise a wide variety of subhabitats from permanent swamp to grassland inundated for perhaps only a few months and thus support particularly diverse animal and plant communities. They are important components, for example, of most of the world's great tropical national parks and not a few of the temperate ones also. A particularly fascinating example is that of the Amazon floodplain (Goulding, 1980) where the flooded forest in the wet season provides, in fruits, seeds, falling insects and the faeces of canopy-dwelling monkeys, a food source for migratory fish in the otherwise nutrient-poor waters. Predators like the spectacled cayman, turtles and piscivorous fish follow the fruit and seed-eating fish into the forest and, in the generally still waters of the forest at high flood, nutrients released into the waters through excretion support a plankton community which in turn supports the young-of-the-year fish. Successful recruitment would not be possible were the fish confined to the main channel of the river to which they must return in the dry season.

Thus, particular wetlands can be seen occupying individual positions in a continuum varying among the important axes of latitude, ombrotrophy and minerotrophy, salinity and hydrology and perhaps among others, perhaps less dominant. Imposed on this constellation are the varied roles wetlands provide for human societies and the problems they may pose for human aspirations.

9.3 HUMAN USES OF WETLANDS

The commodities proffered by wetland habitats to humans can be grouped into three categories: intrinsic values, subsistence values, and commercial values. Intrinsic values are those which accrue by the very existence of the wetland. For example, because of the inevitable waterlogged state of wetlands and the low solubility of oxygen in water, decomposition of litter is hampered and peat deposits frequently build up. These represent a net removal of carbon from the carbon cycle and a net addition through release of oxygen in photosynthesis. The relative importance of this in maintenance of a particular atmospheric concentration of oxygen may be great (Lovelock, 1979) and the changes in atmospheric carbon dioxide content from the burning of oil and coal (fossil wetland deposits) is well established (Bolin *et al.*, 1986). The drainage and peat oxidation of existing wetlands may also have made a significant contribution to increased atmospheric carbon dioxide levels. Methanogenesis from peat soils is also a potentially important regulator of atmospheric composition (Fowler, 1990).

A further intrinsic value results from the role of wetlands in the hydrological cycle. The impedance caused by their biomass and structure and the absorptive properties of peat and particular genera like *Sphagnum* serve to store water in floods and hence to buffer the downstream effects of storms (Novitzki, 1982). Such reservoirs of water may also allow recharge of local groundwater aquifers and encourage sideways flooding of grasslands to the fringes of permanent swamps in a valley. Such floodplain systems in tropical countries are extremely fertile and rich in wildlife because of the silt distributed by the flood and the variety of subhabitats created (Wellcome, 1979; IUCN, 1984). Drainage of swamp systems results in transmission of flood effects downstream and the need for engineering defences to protect farmland and property, much of which has itself been unwisely established on the outer parts of the natural river bed: the floodplain.

A third intrinsic value is that of the protection that wetlands provide to coasts fronted by salt marsh or mangrove. The wave energy that can be absorbed by the multiple flexible stems of a wide marsh or swamp is much greater than that which a concrete or earthfill structure can take without damage (Teal and Teal, 1969). Fourth, because of their complexity and the abundance of water, many wetland systems are of key importance for their wildlife, both in aesthetic terms and in the material benefits exploitation of their gene pools might confer. Rice (*Oryza sativa*), the crop grown most abundantly on a world basis, is derived from wetland ancestors, and with changes now occurring in the biosphere at unprecedented rates, it is likely that this resource might need to be exploited to a large extent by plant breeders. Furthermore, it seems clear that many coastal wetlands are important nurseries or feeding areas for offshore fish, some of them commercially exploited, and that organic detritus from the wetlands may be crucial to offshore foodwebs.

The second group of commodities includes subsistence values. Even in far northern regions groups of people have traditionally depended on wetland systems for food, building and other raw materials (Dugan, 1988). Some of them, the Madan in Iraq, for example, subsist in permanent swamps (Thesiger, 1964). The Madan people fish, graze water buffalo in a marsh dominated by *Phragmites* and build complex reed buildings on platforms also built from piles of reed.

Other groups have annual cycles of transhumance dependent on the rise and fall of river levels in floodplain systems. The pattern is often one of crop growing on the upland or on parts of the floodplain grassland at the height of the flood or as it recedes. This is followed by cattle, sheep or goat grazing as the flood further recedes and a vigorous flush of grass is produced. Finally, in the difficult period of the dry season the permanent swamp along the central river channel may provide a refuge of fish and plant rhizomes. Examples have been recorded in Africa along the Nile and rivers in both the west and south of the continent and in western India. However, there are probably few examples of tribal societies that do not have some dependence on wetland systems because of the key importance of a freshwater supply to all peoples.

There is no definite boundary between subsistence and the third category of commodity exploitation on a commercial basis. Many subsistence peoples now use wetland products to barter for products from elsewhere and may even use them in a cash economy. The Madan, for example, sell fish, reed mats and rice outside their own societies. And the seventeenth to nineteenth-century communities of the East Anglian fenlands and Broadland sold marsh hay, wildfowl and fish to markets as far away as London (Ellis, 1965). In Bangladesh a sustainable industry based on rattan, timber and honey is carried on in the mangrove swamps. Inshore and shellfisheries and fish culture are commercial uses frequently made of far-eastern mangrove swamps.

Recreation also falls into the continuum between subsistence use and commercial exploitation. It may provide some income to local peoples associated with national parks and nature reserves as wages for game rangers, and those associated with accommodation and catering. Yet in other circumstances it may be damaging to the system if it involves major disturbance by the building of roads or excessive provision of boats and facilities. In some circumstances a degree of commercial exploitation through traditional methods of management may help to maintain the diversity of a wetland. In the fens of the Norfolk Broadland, for example, the cutting of reed for sale as thatch halts succession to alderwood and creates a mosaic of habitats albeit not quite so diverse as they were before mechanized methods of harvesting were employed (Nature Conservancy, 1965).

It is, however, in the three overt forms of commercialization that great damage and sometimes complete obliteration of wetlands occurs. These are permanent flooding behind dams created for irrigation storage or hydroelectric power (Goldsmith and Hildyard, 1985), the mining of peat for fuel or horticulture (Moore and Bellamy, 1974), and the engineering operations of embankment, infill and drainage which convert wetlands to dryland for farming, urban or industrial development (Shoard, 1980).

Dam building was an activity perceived to create prestige for the governments of developing countries in Africa in particular during the 1950s and 1960s when the Kariba, Akosombo and Aswan High Dams were created on the Zambezi, Volta and Nile respectively. Many smaller dams have been created elsewhere in the continent, and in India and elsewhere. The impetus has now moved to Brazil and China with plans now laid for schemes of huge scope (Caufield, 1985a, b). The balance of advantage and disadvantage in dam building has always been controversial. The cost-benefit analyses drawn up for the schemes tend not to include the array of environmental disbenefits. These include displacement of indigenous peoples from the floodplain and loss of the resources of the plain, downstream effects on fisheries and in the case of the Nile on the economy of the Nile delta, rapid infill of the dam owing to high sedimentation rates from catchments deforested as part of the general development of the area, salinization of lands irrigated from the reservoir, disruption of communications and the increased spread of water-borne disease. Traditionally the impetus for dam building comes either from a civil engineering lobby with little vision beyond its own construction sites or from governments with scant regard for the rural sections of their population. Not surprisingly a vociferous lobby amply backed by information from the consequences of past schemes now fails to see any net advantage in most new schemes.

Some countries with extensive blanket bogs and muskeg mine peat on a large scale to burn in their power stations (Moore and Bellamy, 1974). The former USSR, Canada and Ireland are examples. In other countries (particularly in the UK and the Netherlands), much smaller bogs are mined for peat for horticultural use. In the UK the statistics of loss of lowland peat mines are startling (Nature Conservancy Council, 1984) with the great majority of bogs that were extant in the 1940s now so very seriously damaged that conservation agencies in the 1980s have been obliged to mount campaigns to 'Save our Peatlands'; the peat extraction industry has countered with its own propaganda. Peat cutting has been a traditional industry for centuries in Western Europe and hand cutting was a slow enough and delicate enough operation not to damage the bog overall. The site was not drained and peat began to re-form in the cuttings which to some extent provided a variety of habitats on the bog surface. The Norfolk Broads in East Anglia are a set of lakes formed in mediaeval peat cuttings that have the particular charm resulting from traditional methods of exploitation which, perhaps as much by accident as design, cause no permanent damage (Ellis, 1965). Commercial cutting, on the other hand, generally involves extensive damage and scarification by machine which leaves a raw, bare surface that is dry and dusty and eventually colonized by a scrub lacking the former bog plants and preventing any further peat formation. Peat mining on this scale has the same characteristics of other sorts of fossil fuel extraction. It steadily uses up a resource which is replaceable only on a geological time scale.

The statistics of damage to wetlands through drainage and reclamation are equally dramatic. In the UK the great area of fenland surrounding the Wash suffered its first major drainage in the seventeenth century although there had been several attempts from the Roman occupation onwards. The fenland has now essentially been lost to high intensity agriculture and although ostensibly it produces a rich crop of wheat, sugar beet, vegetables and cut flowers, it has been at the cost of maintenance of a flood defence system and a system of agricultural subsidies that no one has yet dared to balance in a properly based cost-benefit analysis.

In general the encouragement by the 1947 Agriculture Act and the subsequent systems of advice and subsidy to maximize agricultural production in the UK following the World War II have had a catastrophic effect on the UK wetlands. Extensive areas of riverine water meadow, grazing marsh and wet pasture have been drained and even small pockets of marsh at the corners of meadows have been filled in so as not to impede the operations of large machinery. The former richness of Romney Marsh, the Pevensey Levels, the Broadland grazing marshes, the Somerset levels, the Lancashire Coastal Plain and the Derwent Ings has been near obliterated as part of an overall trend in agriculture that was unforeseen 40 years ago but which has been devastating (Shoard, 1980; Green, 1985). To add to the effects of agribusiness there has been the steady loss of sites for municipal rubbish tipping, the reclamation of saltmarsh for agricultural and of mudflat in estuaries for industrial development (Shaw, 1983).

On a world scale these UK trends have been mirrored, with the Worldwide Fund for Nature and the International Union for the Conservation of Nature seeing the need to mount their own wetlands campaign following the identification in the World Conservation Strategy (1980) of wetlands together with forests and prime agricultural land as a group of particularly threatened habitats. The losses of mangrove swamp to the development of marinas, housing, salt evaporation and industry have been particularly large.

9.4 VULNERABILITY OF WETLANDS

Wetlands have suffered damage at the hands of the developed and developing worlds more than most ecosystems. Although open fresh-waters and the open sea have very positive images in the public psyche, associated with romantic legends, tales of brave voyages, perceptions of escape from hum-drum lives and a sense of purity and clarity, wetlands evoke many opposite responses. Muddy, boggy, smelly, mired-down, sucked down, marsh fever, ague, and leeches are some of the negative associations. Although we may have had an aquatic phase in our past (Hardy, 1960), we are not water creatures and our current urbanization separates us more and more from the reality of all natural situations. Our sallies onto open water are mostly in floating chunks of the city, and second-hand information seems to strengthen our fears and misconceptions. Among the general public there is thus relatively little support and understanding when a defence must be mounted to save a marsh or oppose exploitation of a bogland compared with that which can be mustered for a forest or downland. The agricultural and industrial lobbies in the past have thus found it quite easy to manipulate opinion in favour of the developments they have planned, and in the UK have managed to retain total control of the internal Drainage Boards without difficulty. Wetland has very definitely been wasteland (Baldock, 1984).

The negative associations that wetlands evoke may come from historical associations. Marshes were difficult places to cross as evinced by the Neolithic and Bronze Age wooden trackways built over the peat in Somerset (Godwin, 1978; Coles and Orme, 1982) and the East Anglian fenland, the use of islands among such wetlands for villages that could be readily defended, and the selection of islands in the fens for the building of secluded monasteries where the great cathedrals of Peterborough and Ely now stand. Marshes were also unhealthy places where death rates from malaria in the UK were comparatively high: indeed a considerable number of parish clergy refused to live near them, giving as reasons to their bishop the unhealthiness of the marshland parishes (Dobson, 1980).

The association with disease is well taken still, because of the diseases that plague huge numbers of the world's population, it is those with wetland vectors that are among the most common. Malaria infects 250 million people at a time and kills 100 000 per year; for schistosomiasis and filariasis there are more than 200 million and more than 300 million cases respectively (Peters and Gilles, 1977). A diverse menagerie of other flukes, tapeworms and other nematodes infests considerable numbers.

Other diseases like typhoid, polio, cholera, hepatitis and infant diarrhoea, although they are essentially the diseases of dense populations poorly served by public health measures, have also become associated with water for that is the medium by which they are usually transmitted.

A second reason for the vulnerability of wetlands is their fertility. Materials collect in basins and in the case of wetlands these include fertile silts and organic soils. Wetlands often have a very high primary productivity because of this fertility, coupled with the availability of water and not surprisingly they can be converted to fertile farmland if drained to an adequate degree. The fertility may not be sustained but the time-scale for its loss is decades or centuries, so its decline is not realized in the lifetimes of the drainers. Peat soils oxidize on drainage and in all cases the land sinks, thus needing even greater flood protection. Drained peat soils may also, through oxidation of iron sulphide deposited within them from the activities of sulphate-reducing bacteria in the anaerobic wetland soils, produce sulphuric acid and deposits of ochre. The latter is a complex of iron oxides and hydroxides which is unsightly and potentially damaging to the watercourses to which the peat is drained. Fertility also declines because the silt supply formerly deposited in the wetland is now channelled through the drainage system and a need for artificial fertilization soon develops if the land is cropped.

The third reason for vulnerability is simple ignorance on the part of most people on environmental matters (Grieg *et al.*, 1989). We have become separated from our natural environment and tend to see ourselves and our cities as self-contained islands in a sea of surroundings that has little consequence. Technology has made life very comfortable for most in the developed world. Our concerns have become the concerns of self. We have been manipulated to expect that governmental and industrial institutions will maintain our comforts and to accept that their interests are our interests. Even worse, propaganda makes us believe that our apparent wealth comes from commerce and industry; the reality is that it comes ultimately from photosynthesis and rocks. Our connectedness with the intrinsic understanding of the operation of the global biosphere have been undermined. For all our information and technology we are now truly dwarfs compared with those who keep body and soul together in the Kalahari desert or the Nile sudd.

9.5 SOLUTIONS BY TINKERING

The fate for all wetlands is ultimately destruction for good reasons or conservation. The uncertainty lies in what constitutes 'good reasons'. There are in the rich Western world no longer any positive reasons for the further destruction of wetlands. Agriculture is already sufficiently efficient to provide more food than needed, especially when the wastes of processing to provide goods marketed to artificially manipulated standards are considered, and public health measures already control water-borne diseases without the need for destruction of the vector's habitat. Clearly also, the interference with wetlands for flood control has led to contingent and expensive need for further engineering elsewhere. Much of the existing damage has resulted from honest mistakes as much as from uncontrolled avarice but there is no longer any excuse for not altering legislation that permits continuation of the latter. In the UK there is little protection for natural and semi-natural habitats. The designation of Site of Special Scientific Importance status gives far from absolute protection and the recent dismemberment of the UK Nature Conservancy Council (Environment Protection Act, 1990) promotes little confidence in Government commitment to environmental matters.

There is a fatal flaw in the legislation of all Western countries in that the management of rivers, wetlands and lake systems is divorced from the management of the catchments from which the water is derived. Thus in the UK the National Rivers Authority falls under the aegis of the Department of the Environment while agricultural policy is determined by the Ministry of Agriculture, with powerfully established existing lobbies. It has been clear to professional ecologists for some time that the only valid unit of study is the catchment system not the water-containing basin, which represents only the lower part of an overall unit. The Water Act of 1973 at least gave to the Water Authorities it established control over the waterways of entire catchments but no influence over the management of the catchment itself; the Control of Pollution Act 1974 specifically excludes normal agricultural operations from prosecution even if they cause significant damage through siltation and eutrophication to receiving waters. The Water Act of 1989 contains provision for protection zones but no case lore yet exists to establish how effective the provisions will be. Overall the approach to conservation and management of wetlands in the UK can be described therefore as piecemeal and uncoordinated. Even when private bodies such as the County Wildlife Trusts are able to purchase threatened wetland areas, they are frequently unable to manage them effectively for lack of control over the quality of the water they receive.

For developed countries like the UK there is no real dilemma. There is merely a dual moral requirement like that which the health authorities of any civilized society have, to ensure the health of the few, highly valuable survivors of a rife plague and at the same time to remove the causes of the plague. Even then with our present value system, the solution is one of tinkering rather than of fundamental understanding.

In the developing world, the situation is more complex and although the approach is equally piecemeal the solutions are likely to be more difficult to implement. Some observers in the Western world would point to a package of Third World problems whose solutions include the need for irrigation, power supplies and living space in floodplains close to sources of freshwater for domestic use. This package usually focuses on increasing population sizes and an assumption that a degree of organization and enterprise along western lines is what is required.

This view underlies the damage that has been done to wetland and other ecological systems in the developed world and is clearly mistaken. It assumes that the lot of all but technological man must be nasty, brutish and short, and that nothing may be learned from cultures other than that which has characterized north temperate regions for a few hundred years. We should remember past and powerful city states whose demise was strongly linked with overexploitation of their local resources. Rome is perhaps the best example (Seymour and Girardet, 1988). There are many examples of peoples who through close dependence on their environment have developed harmonious ways of exploiting it without destruction (Reader, 1988). It behoves us to study these examples very carefully. They embody principles which we in the Western world are going to have to return to in the near future as the increasingly tangled web of problems into which our technology leads us, moves us inexorably towards a violent competition for natural resources which are dwindling and irreplaceable. In the long term we will be forced to abandon the systems which have led to the destruction of much of our resource base, of which wetlands are but one example, and to develop systems which have a more fundamental basis than that of individual acquisition. The present solution to wetland problems is piecemeal – a nature reserve there, a national park here – and is a product of the contemporary fashion for reductionist thinking. Such solutions have no future.

REFERENCES

Baldock, D. (1984) *Wetland drainage in Europe*, IIED/IEEP, London.

Bellamy, D. (1986) *The Wild Boglands*, Helm, London.

Bolin, B., Doos, B.R., Jager, J. *et al.* (1986) *The Greenhouse Effect, Climatic Change and Ecosystems*, SCOPE 29, J. Wiley, Chichester.

Caufield, C. (1985a) *In the Rainforest*, Heinemann, London.

Caufield, C. (1985b) The Yangtze beckons the Yankee dollar. *New Scientist*, 5 December 1955, **108**, 26–7.

Coles, J.M. and Orme, B.J. (1982) *Prehistory of the Somerset Levels*, Somerset Levels Project, Cambridge.

Dobson, M. (1980) Marsh-fever – the geography of malaria in England. *J. Hist. Geog.*, **6**, 357–89.

Douglas, M.S. (1947) *The Everglades: River of Grass*, Ballantine, New York.

Dugan, P.J. (1988) The importance of rural communities in wetlands conservation and development, in *The Ecology and Management of Wetlands* (ed. D.D. Hook), vol. 2, Croom Helm, London, pp. 3–11.

Ellis, E.A. (1965) *The Broads*, Collins, London.
Fowler, D. (1990) Methane, ozone, nitrous oxide and chlorofluorocarbons, in *The Greenhouse Effect and Terrestrial Ecosystems of the UK* (eds M.D. Hooper, and M.G.R. Cannell), HMSO, London, pp. 10–13.
Godwin, H. (1975) *The History of the British Flora*, Cambridge University Press, Cambridge.
Godwin, H. (1978) *Fenland: its Ancient Past and Uncertain Future*, Cambridge University Press, Cambridge.
Goldsmith, E. and Hildyard, N. (1985) *The Social and Environmental Effects of Large Dams, 1, Overview*, Wadebridge Ecological Centre, Wadebridge.
Goulding, M. (1980) *The Fishes and the Forest: Explorations in Amazonian Natural History*, University of California Press, Los Angeles.
Green, B. (1985) *Countryside Conservation*, Unwin, London.
Grieg, S., Pike, G. and Selby, D. (1989) *Greenprints for Changing Schools*, Worldwide Fund for Nature and Kogan Page, London.
Hardy, A. (1960) Was man more aquatic in the past? *New Scientist*, **7**, 642–5.
International Union for the Conservation of Nature (1984) *Directory of Wetlands of International Importance*, IUCN, Groningen.
International Union for the Conservation of Nature and United Nations Environmental Programme and Worldwide Fund for Nature (1985) *World Conservation Strategy*, Gland, Switzerland.
Lovelock, J. (1979) *Gaia. A New Look at Life on Earth*, Oxford University Press, Oxford.
Moore, P.D. and Bellamy, D.J. (1974) *Peatlands*, Elek, London.
Nature Conservancy (1965) *Report on Broadland*, HMSO, London.
Nature Conservancy Council (1984) *Nature Conservation in Great Britain*, NCC, Peterborough.
Novitzki, R.P. (1982) *Hydrology of Wisconsin Wetlands*, US Geological Survey Information Circular 40, pp. 1–22.
Peters, W. and Gilles, H.M. (1977) *A Colour Atlas of Tropical Medicine and Parasitology*, Wolfe Medical, London.
Reader, J. (1988) *Man on Earth*, Collins, London.
Seymour, J. and Girardet, H. (1988) *Far from Paradise*, Green Print, Basingstoke.
Shaw, D.F. (1983) Conservation and development of marine and coastal resources in *The Conservation and Development Programme for the UK* (ed. F. O'Connor), Kogan Page, London, pp. 262–312.
Shoard, M. (1980) *The Theft of the Countryside*, Temple Smith, London.
Stead, I.M., Bourke, J.B. and Brothwell, D. (eds) (1986) *Lindow Man: The Body in the Bog*, British Museum Publications, London.
Teal, J. and Teal, M. (1969) *Life and Death of the Salt Marsh*, Little, Brown & Co. Boston.
Thesiger, W. (1964) *The Marsh Arabs*, Longmans, London.
Wellcomme, R.L. (1979) *Fisheries Ecology of Floodplain Rivers*, Longman, London.
Wheeler, B.D. (1980) Plant communities of rich-fen systems in England and Wales. I. Introduction. Tall sedge and reed communities. *J. Ecol.*, **68**, 365–95.

Case study: nature conservation – a Scottish memoir 10

J. Morton Boyd

John Morton Boyd has spent most of his working life in his native Scotland, where he was Director of the Nature Conservancy Council for 14 years. In the 1960s he was a Nuffield Travelling Fellow in the Mid-East and East Africa. In the 1970s he played a prominent part in the Anglo-Soviet Joint Environmental Commission, and the Commission on Ecology of the International Union for the Conservation of Nature (IUCN). He is now conservation consultant to the Forestry Commission, Scottish Hydro-Electric plc, and the National Trust for Scotland. His books include The Highlands and Islands *(with Frank Fraser Darling),* The Hebrides: a Natural History *(with his son, Ian Boyd), both in the Collins New Naturalist Series, and* Fraser Darling in Africa: a rhino in the whistling thorn.

10.1 THE BIRTH OF AN IDEA

The Loch Lomond road runs tortuously northward from the Clydeside conurbation into the Scottish Highlands. Immortalized in song, the collective memory of the Highlands in many Scots is charged with the beauty of that drive deeper and deeper into the mountains. Lochside woods are resplendent in seasonal garb with vistas of tree-clad islets and distant peaks. Hamlets and cottages become fewer on the journey, and

Environmental Dilemmas Ethics and decisions
Edited by R.J. Berry
Published in 1992 by Chapman & Hall, London. ISBN 0 412 39800 1

the far side of the loch bears little obvious sign of human influence. There is a distinct sense of having passed through a gateway from the man-made to the natural environment. The wide, spacious loch of the south becomes a narrow, profound fjord in the north, in which all sight of the Lowlands is lost. Loch Lomond is an important part of Scotland's heritage and its designation as a national park has remained an unresolved issue for almost half a century.

There is one big surprise. Towards the head of the loch at Inveruglas the natural, rural aspect of the drive is interrupted by a large hydroelectric power station to which four large pipes descend the slopes from a tunnel through the core of Ben Vorlich. These convey water from Loch Sloy high in the mountains to the west. Although great trouble was taken to minimize the intrusion of this industrial installation into the wonderfully picturesque rough bounds of Arrochar, it is still a cultural shock. If such a scheme were mooted today there would doubtless be a national furore. As it was, the scheme was conceived during the years of industrial boom in the late 1930s and given great impetus by the immense war effort of Clydeside. Heavy industry required power to meet daily peak load, and although Loch Sloy had water enough for only 2 hours of daily running, the fully operational station could provide 130 MW in that time. The founding fathers of the Scottish hydroelectricity would be amazed at the huge quantities of electricity that are available today, and of how comparatively small is the contribution of Loch Sloy; the new nuclear station at Torness in East Lothian, if fully operational, can supply 1320 MW for 24 hours daily.

The opening of the Loch Sloy scheme in 1950 was the realization of the dreams of engineers who saw power in the Scottish lochs and rivers, and of politicians who saw *that* power rehabilitating the Highlands and Islands. 'It means something more than millions of kilowatt hours, or thousands of tons of concrete and steel; it crystallizes the imagination, enterprise and effort behind the great development . . . which has as its aims improved standards of living, more employment and increased production . . . (showing) the new spirit that exists where before there was only depopulation and despair', were the words used by Secretary of State Thomas Johnston at the opening (Payne, 1988). As Scottish Secretary, he had created the North of Scotland Hydro-Electric Board. He also discovered and promoted the pioneer ecologist Frank Fraser Darling who, from the standpoint of a conservationist working in the West Highlands, shared Johnston's vision of hydroelectricity as the power behind an integrated industry of agriculture, forestry, fisheries and tourism. Sadly, neither of them foresaw the consequences of industrial decline in the UK, and the collapse of their dreams of a largely self-sustaining Highlands.

If the end of the war was a time of new beginnings for industrial development of which hydroelectricity was only a symbol, it was also the birth of the modern environmental movement in the UK. The two issues were not unrelated. Following the uncivilized trauma of war, these new beginnings were tempered with a civilizing influence in the appreciation and use of natural resources. As a young engineer working on the construction of Loch Sloy pipelines I was aware of this. I admired greatly the big-hearted spirit of my engineering mentors in overcoming the forces of nature. I saw an intrinsic beauty in the precision of their technology and in their works in concrete and steel. To the technical eye, the scheme was an epic in the harnessing and unleasing of the might of nature. There was no conflict in the engineers' minds about the probity of their actions against the serene backdrop of Loch Lomondside. Their work was professional, honest and even patriotic.

I had greater difficulty in reconciling that probity. Although I did not know it at the time, the benign influence of wilderness which so gripped John Muir in the high Sierra of California, had touched me in my own way in my own place. From the human vulgarity of the workers' camp to the blasting open of the mountain, I saw a gigantic offence on nature which would endure for all time, graven on the face of the mountain.

To escape from the harsher realities of life on the job, I took a tent to the lochside where the call of sandpipers and the balm of lapping waters replaced the racket of the camp. Weekends and holidays I spent in the hills with deer and ptarmigan, far from pneumatic drills and rivet guns. Was mountaineering simply a thirst for adventure following my flying days in the RAF, or was I running away from my destiny? I came from a line of stonemasons and blacksmiths of which I was proud. Engineering was bred into me, but I was possessed by a new and compelling idea, as much mystical as it was rational: a burning desire to research nature in order to communicate its beauty to my peers, and to care for it had become an imperative in my life. Although I did not know it at the time, the personal dilemma from which I emerged in 1948 to follow a career in conservation instead of engineering, was to become in later years, one of global proportions. Today, it is enshrined in the concept of 'sustainable development', which has become a slogan in world conservation, a centre piece of international congresses on the environment, and a plank in the policies of national governments.

10.2 CONSERVATION: A MATTER OF NO COMPROMISE?

I had not travelled far in my new career before the realities of my decision were borne down upon me. Most of the UK – and the world – had been changed irrevocably by man's use of natural resources. The entire face of the countryside had been changed, and the vast inventory of living creatures bestowed upon the British Isles by nature since the last ice age had been significantly depleted. Natural glacial landforms had been altered out of all recognition. Naturally wooded landscapes had been transformed by centuries of agriculture, animal husbandry and fire. Little or no thought had been given to the conservation of species, habitats, geological and physiographical features. I began to feel that conservation was still held at the starting gate of the race in human destiny, with development already in the home straight. I was imbued with a mixed sense of grievance, urgency and determination. No time, I thought, should be lost to make up ground, and that further losses to the inventory be resisted with all my strength.

Later, when I worked in Africa and the Middle East, I came face to face with a paradox. As a conservation evangelist, I was there to help the people of the Third World to save their wildlife by conserving *inter alia* their natural forests and scrublands. Yet we in the UK had lost all but a few fragments of our natural woodlands. Many of the East African forest officers had been trained in Aberdeen under Professor H.M. Stevens and they knew well the sad history of the native Caledonian pine forest, on which Stevens was the authority. The scale is different, but in principle and as a lesson in conservation, the clearance of Scottish forests by people over ten centuries, together with the clearance of the people from the land in the eighteenth and nineteenth centuries, bears comparison with the excesses of the present day clearance of the tropical forests. My blushes were spared, however, as these enlightened men did not doubt the sincerity of my mission, springing as it did from a desire to avoid a similar tragedy being visited on their land and people.

I was conscious of, and greatly inspired by, the roots of conservation in the UK, which were different from other countries. The first moves for conservation *per se* sprang from John Muir's America in the second half of last century, supported by philosophers like Emerson and Thoreau who were occupied with a specifically American attitude to nature. Thoreau's well-known quotation 'in wildness is the preservation of the world', became an article of faith in conservation, espoused later by Theodore Roosevelt and Aldo Leopold. In the UK, the great stimulation given to natural science by Darwin led to a greatly enhanced knowledge of, and interest in, wild creatures, and it was not long until their *preservation* became an issue, promoted by the naturalists A.A.R. Horwood, F.W. Oliver, and the Honorable N.C. Rothschild, who founded the Society for the Promotion of Nature Reserves in 1912. Later the cause was translated into one of *conservation* supported by the British Ecological Society with A.G. Tansley, C.S. Elton, Julian Huxley, and W.H. Pearsall in the lead. The whole story is told by Max Nicholson in *The Environmental Revolution* (1970) and John Sheail in *Seventy Years in Ecology – The British Ecological Society* (1987), and shows the conservation movement in the UK rooted in the animal and plant protection movement and academia. It flowered in the Nature Conservancy, a government agency appointed by Royal Charter under the Privy Council, and possessing an independent voice proud of the large government departments. By contrast, nature conservation in most countries on both sides of what was the Iron Curtain, has grown out of departments of government – lands, forestry, national parks, game, fisheries and wildlife, and occasionally tourism – and is largely controlled by them.

Although the independent voice and powers of compulsory purchase possessed by the young Nature Conservancy were used with the utmost care and discretion, they served to nerve the arm of conservation in its confrontation with development. In 1958, a year after I joined the Conservancy, there was a 'David and Goliath' contest between it and the mighty Central Electricity Generating Board over the nuclear power station at Dungeness. No one expected the small, almost unknown Nature Conservancy headed by its Director-General Max Nicholson to make an impressive showing. The CEGB won, but the Conservancy emerged with a fearlessness of industrial giants that has endured ever since. Later, in the mid-1960s, a Parliamentary battle took place, promoted by the Tees Valley and Cleveland Water Board. It was a proposal by Imperial Chemical Industries for the construction on Teeside of two gigantic ammonia plants. The impact on the vegetation of Cow Green, in the Upper Teesdale National Nature Reserve, involved the drowning of about 20 acres of sugar limestone, bearing rich and rare plant communities. A Teesdale Defence Committee, composed of the country's top botanists, supported by the British Ecological Society, defended Cow Green. The battle, which started as far back as 1956, was not concluded until 1971 (Sheail, 1981). The conservation case was lost, but the Cow Green debate served to demonstrate to MPs a previously unknown force in their midst, and a new type of political dilemma.

Opinion was divided as to the function of the Nature Conservancy which in 1973 became the Nature Conservancy Council (NCC). One side believed that it should advise strictly on the scientific aspects of conservation, and leave political decisions entirely to ministers; the other side saw the Conservancy balancing scientific advice with political reality, and thus relieving ministers of some onus of decision. The NCC was not an elected body, and although given an absolute power of decision in matters strictly scientific, it was answerable to Parliament through ministers. This dichotomy of view lead to much confusion and ill-feeling in the conservation movement, particularly as the larger voluntary bodies became active in the political lobby. The Council and Committees of the Conservancy included landowners, farmers, industrialists, media persons, and politicians as well as natural scientists, and their discussions became inescapably political. The crux of the matter lay in the statutory duty laid upon the Conservancy to notify Sites of Special Scientific

Interest (SSSI) to the local planning authority. This was a duty set by Parliament and not by Government, so no minister could intervene if 'in the opinion of the Conservancy' a site should be notified. As such a notification could affect the value of the land, the lack of any appeal procedure against notification was widely criticized and featured prominently the debate on the Scottish Natural Heritage Bill.

Thus the world-wide confrontation between conservation and development was mirrored in the UK, highly tempered here by different perceptions and compromise. However, much of the heat has been taken out of the issue by the provisions made in the Wildlife and Countryside Acts (1981 and 1985) for compensation to farmers and foresters for legitimate profit foregone through a management agreement with the Nature Conservancy Council (NCC)* on the notified SSSI. None the less, there has since been a number of outstanding test cases in Scotland – Lurcher's Gully on Cairngorm (ski development), Duich Moss in Islay (peat extraction), and the Flow Country in Caithness (afforestation of mires), Glen Lochay in Tayside (afforestation of species-rich upland – and the SSSI procedures still give cause for concern. The Scottish Land Tribunal awarded the owner of the Glen Lochay estate a sum of £568 000 plus interest and costs as compensation for profit foregone on 3240 ha, in the interests of nature conservation. The media made the mistake in seeing this as a victory for the owner; in fact, it was a substantial, if expensive, gain for conservation on the Alpine habitat of the Breadalbane Hills, which otherwise would have been clad in coniferous forest. It is an irony of history that the owner, described by *The Scotsman* as 'the biggest sheep farmer in Europe' – the modern counterpart of the flockmasters of the past 200 years who, with sheep and fire stick, have prevented the natural regeneration of woodlands in the glens – should have been compensated from public funds for *not* planting trees.

The dilemma of whether to develop in the face of a conservation restriction is political and not scientific. Following enormous work on reviews of nature and geological conservation (and in train for marine conservation), the scientific framework has been established. In scientific terms it is challengable only by competent, peer-reviewed work of similar scale. On the other hand, the rules of conservation management on the ground as they affect the choice of boundaries and the effects of management, are nothing like so well worked out. This is the flash point of compromise which requires much insight into the function of natural systems, and the predictable reactions of the negotiating parties. Over the years of SSSI casework, I have felt the pressures of no compromise upon me. Arising spontaneously and hardened within by my evangelism, not

* The functions of NCC are now vested in Scottish Natural Heritage, English Nature, the Countryside Council for Wales, and the Joint Nature Conservation Council.

to mention the occasional truculence and disparagement of the contesting party, the spirit of compromise was engendered by imaginative and business-like discussion resulting in no damaging loss on either side. Today there are over five thousand SSSIs in Great Britain. Although many still suffer loss and damage and a few reach the national headlines, the vast majority work well. In *The New Environmental Age,* my mentor Max Nicholson (1987) provides the vision of reconciliation between conservation and development:

> Our aim must be to bring about a situation where . . . in all humanity and its institutions, . . . promoters of environmental conservation as a separate entity will be found superfluous, and will gently and thankfully fade away, greeting with reverence and appreciation the new environmental age, which can be discerned only with the eye of faith.

10.3 DUCKS VERSUS JOBS

I had just ascended the last 'pitch' of my climb to the top of the NCC in Scotland, when there broke upon me and my team the tidal wave of North Sea Oil and Gas (Cairns and Rogers, 1981). Nature conservation as a statutory element of planning control was widely misunderstood, sometimes ignored, and frequently unknown to the developers. Many of them were foreign: American, French, Norwegian. Local authority planners were well enough versed in the SSSI scheme to notify us somewhat timorously of proposed developments. This notification was not given in time. We were informed only when the developers had invested huge sums of money in the survey and technical confirmation of the sites for a variety of installations, such as oil rig and production platform construction yards, oil and gas terminals, oil ports, tank farms, and servicing bases. This was often accompanied by the unveiling of the development to the media, with an application for planning permission *in principle*. There was great irritation and annoyance among developers and local councillors when the development infringed a Site of Special Scientific Interest (SSSI), which were abundant in sensitive coastal areas. To be fair to local planning officers, some of these SSSIs had not yet been properly notified; the machinery could not be set up quickly enough to deal with the on-rush of development.

Rocky peat-clad islands in Orkney and Shetland were suddenly mooted for vast oil terminals; spacious sand dunes and tidal flats in the Moray and Cromarty Firths were to have huge graving docks and steel erection sites; an oil pipeline, buried in its length through the Scottish countryside, was to be constructed from Cruden Bay to Grangemouth, and a gas pipeline south from St Fergus to the national grid. Outstanding scenic and wildlife areas such as Loch Kishorn and Loch Fyne were developed and others, Drumbuie, Loch Eribol, Dunnet Bay and Loch of Strathbeg, were proposed but later abandoned before a sod was lifted. A graving dock, construction yard, and workers' village at Portavadie, Loch Fyne was built but never used.

During 1970–75, 60 000 new jobs were created and 2000 ha of land were developed, holding five major oil and gas terminals, four land pipelines, 15 other major developments, and many servicing bases. Life in 14 different communities was greatly affected (Lyddon, 1981). Aberdeen and Inverness became boom towns. The industrial, social and political impetus in the Highlands and Islands was enormous. The Scottish Industry Department, Highlands and Islands Development Board, and the local councils were wooed by the Industry; there was a planner's vision a 'linear city' fringing the firths from Nairn to Invergordon; in Shetland and Orkney successful accords were struck between the Islands' Councils and the Industry in the development of oil ports at Sullom Voe and Flotta. Rapid progress was an imperative, and any hold-up was costly, seen as a threat to jobs and highly unpopular. 'A chance missed may be a chance lost' said George Younger, the Scottish Secretary. Life was uncertain and would never be the same again.

In 1972 the NCC had just moved into new regional offices in Inverness and Aberdeen, and were both underfunded and undermanned to deal with the holocaust. At the first stage of planning (*'in principle'*), some developers were unexpectedly confronted by an objection from the NCC on the likely damage of the proposed development to the natural environment. The position appeared ludicrous and chimed with Compton Mackenzie's lampoon *Rockets Galore*, in which the grand plan for a guided missile range in the Hebrides was held up because of its likely damage to the local island population of birds. First reactions were to laugh it off, but attitudes changed when the statutory implications became clear, together with the support which environmental protection commanded in Parliament and the wider community.

The atmosphere was frequently tense. We were in the pillory, accused of dragging our coat tails and introducing trivial objections at a time of beckoning prosperity and industrial opportunity. At the time, we did not have the support of the Wildlife and Countryside Act of 1981, which gave much greater strength to site and species conservation, and had to rely much more on persuasion than on binding obligations. The 'green' movement had yet to gain its momentum. Local support for conservation was muted where today it would be much more strident. Sooner or later there was bound to be a major contest.

When the Ross and Cromarty County Council presented its draft structure plan in 1973 the conservation bodies, led by the NCC, objected to proposals for the development of the Cromarty Firth as a major centre for oil and gas development. The spacious tidal flats of Nigg and Udale Bays within the Firth were mooted for partial reclamation and these were an SSSI, one of the most important wintering sites in the UK for waders and sea ducks. A central factor in the conservation case was the extensive areas of the eelgrass (*Zostera* spp.), which is the favourite pasture of wigeon (*Anas penepole*). The decline of wigeon in the UK during this century has been attributed to the disappearance of eelgrass from most of the larger tidal flats (Thom, 1986), and the deliberate removal of a great area of it in the Cromarty Firth was not negotiable. The lines for battle were drawn.

Council chambers in the Highlands and the Islands resounded with a heated 'ducks versus jobs' debate. The media had a field day. The public inquiry at Dingwall attracted national headlines. Both sides threw themselves into the fight with all the skill and force they could muster. In the end the Scottish Secretary, William Ross, was left with the dilemma of political compromise. The Reporter found in favour of the conservation case, but the Secretary, allegedly fearing a nationalist backlash, reversed the Reporter's decision and decided in favour of the County Council's case for development, with a long list of conservation constraints to be observed in the implementation of the plan.

10.4 SUSTAINABLE DEVELOPMENT

The term 'sustainable development' has become a shibboleth of governments and industries, to present a respectful image to a society which is becoming ever more strident in its concern for the environment. It is a concept that was projected onto the world by the Stockholm Conference of 1972, and has been carried ever since by the United Nations Environment Programme (UNEP), the World Conservation Union (IUCN), and the World-Wide Fund for Nature (WWF) in their *World Conservation Strategy* (1980). It has the ring of truth and world-wide acceptance but it is poorly understood by those who use it. There are difficulties in applying it to natural systems, such as primary forest (as compared with, for example, coniferous plantations), and in reconciling its technical and economic implications with the conservation ethic that it enshrines (O'Riordan, 1983).

'Sustainable development' is a theoretical key to improvement of human world order. This is not new but is a restatement of the threadbare 'wise use' concept of the American conservationists of last century, and used by Theodore Roosevelt. Then it was usually accompanied by 'enlightened management', and was meant to convey the notion that, if human beings acted with moderation and restraint in the light of knowledge gained, renewable natural resources such as wildlife would survive indefinitely. However, there is a flaw in 'sustainable development' in nature conservation, arising from the origins of the concept in agriculture, animal husbandry, and forestry. If defined ideally and in isolation, nature conservation is the antithesis of agriculture and forestry. The latter are dedicated to the disciplining of nature, given sanction in the Christian tradition by the divine calling to 'be fruitful, and multiply and replenish the earth, and subdue it: and have dominion over . . . every living thing that moveth upon the earth' (Gen. 1.28).

In this age of 'sustainable development', nature conservation strives to maintain the diversity of life on earth, yet sustainability is dependent on the consistently high quantity and quality of the product. Any threat to sustainability of a marketable product is countered by reducing the threat, usually by fire, or by mechanical or chemical means. This flies in the face of nature conservation with its devotion to the cause of genetical diversity and the survival of species and natural habitats. The development cause is driven by marketing forces such as those of the agrochemical and agro-engineering industries, seeking to expand their business; the conservation cause is driven by the conservation movement campaigning for reduction in the use of chemicals and implements of mass destruction. Positions have become entrenched over decades, the problem of sustainable development in government policy is ubiquitous, and is deepened by the free market which ignores its ethical implications of restraint, moderation, and selflessness. Realising their dilemma, politicians espouse 'sustainable development' as a long-term goal, aware that their own self-interests and those of their constituents are short term, and that they are unlikely to be held accountable for the non-compliance with such a policy.

10.5 'GREENING' OF FORESTRY

In my job at the NCC, I had constant contact with foresters. My first boss, W.J. Eggeling, was an Edinburgh graduate and a distinguished forester in East Africa. Many of my colleagues in nature conservation were forestry graduates. As a natural scientist, I therefore had a somewhat privileged insight into the vocational side of forestry and of how this marched with my own calling in nature conservation. I was conscious that my forestry friends regarded themselves as conservationists in their own right. After all, the Forestry Commission had its internal 'conservancies' and 'conservators' before the inception of official nature conservation. Many of them were practising naturalists, and all of them prided themselves on being countrymen of wide experience and dedication. Wildlife, they said, was an inextricable part of their business, and I required no convincing on that point. I have long recognized that the Forestry Commission, as one of our largest landowners, had more wildlife in its care than any other body and that the Forest Services of the USA and the former USSR vied with each other as the greatest custodians of wildlife in the world.

In the make-up of the forester, however, there was a certain reluctance to be forthright in his support for nature conservation. As I saw it, he (she) was caught in a dilemma: the wildlife of the forest which he secretly espoused, was formally eschewed by his profession. The forestry tradition in the UK springs from the felling of the natural forest over 5000 years and its replacement by agricultural land, livestock pastures, and tree plantations. There is perhaps no woodland left in the UK which is truly natural. There is no ancestral memory in this country of the primaeval forest as there is in those countries which have only lost their natural forests in historical times, or which still possess them.

The collective attitude towards forestry in the UK is essentially exploitation of semi-natural woodland, and sylvicultural progress towards an ideal timber production forest, both at home and in the Empire. It is said that the Indian Forest Service, which sprang from one of the survey divisions of the Indian Army, served as a model for the Forestry Commission when the latter was founded in 1920. The foresters of the first 50 years were greatly influenced by the wartime devastation of the forests, and the 'fortress Britain' concept of self-sufficiency in the case of any further wars. In the depression of the late 1920s and early 1930s the building of this strategic reserve of timber came as a very useful, indeed patriotic, way to provide employment. Thus were planted the great 'tree factories' of Kielder and Thetford, and forestry became an agency of massive change in the countryside, propelled by a production ideal and little else.

Forestry was galvanized by the concept of the strategic reserve and this production ideal. Before World War II the forest-establishment side prevailed but after the war as the forests came to the cropping stage, it was the harvesting and marketing people who had greater influence. They were the celebrants of the production ideal, and in the late 1960s, the strategic reserve concept as the touchstone in forest policy was abandoned and replaced by an array of monetary incentives. Forestry became divided. On the one hand there were forest managers planting new, and replanting felled, woodland; on the other were the harvesters and marketeers who were moved by market forces against the available stocks of saleable timber. What was not widely appreciated at the time was the role which the harvesters played in determining the nature of the future of forests.

Into this scene came the mercurial figure of Dame Sylvia Crowe, a distinguished landscape architect whose teaching of forest design made a deep and lasting impression. Gone were straight forest edges on hillsides, to be replaced with lines flowing in sympathy with skylines and natural features. The landscape architects have since become a permanent implant in forestry and have been joined by ecologists and recreational planners to form a comprehensive environment branch.

When I retired from the NCC in 1985, I was appointed Conservation Adviser to the Forestry Commission. I joined at a time of transition of forestry. Until the early 1980s, the organization was hypnotized by the production ideal and led by the market. Following the Wildlife and Countryside Acts (1981 and 1985) which required the Forestry Commission to take nature conservation into account throughout its operations, the scene began to change. I found that the hard attitudes which I had known in previous times were softening. The personal and corporate commitment to conservation given freedom by the Act, showed itself immediately and grew rapidly. It was a clear case of *glasnost*; many foresters in management positions had for years been awaiting the signal to express themselves freely.

My message was that the foresters' dilemma should be removed. Forestry should no longer be held down to the single purpose of timber production, against its multipurpose potential in the life of the British people (Boyd, 1987). I found that landscape design and nature conservation were described in strategic plans as 'constraints' on forestry, and pointed out that these were inalienable parts of modern forestry, citizens of the same country as 'timber', and deserving a status in their own right. I described forestry as changing from a narrow economic concept to a wider philosophical ideology in which it is seen as a civilizing influence in the way of life and spirit of the British people. I became convinced that forestry in the right hands is gentle, cultured, and spiritually uplifting; in the wrong hands it is harsh, uncouth, and depressing.

I take great satisfaction in the thought that in 1991 the Forestry Commission adopted a multipurpose definition for forestry similar to that which I proposed in 1985. My efforts have not been in vain.

10.6 COMING FULL CIRCLE

It is an irony in my life that after 40 years in nature conservation I should find myself back in hydroelectricity as the conservation adviser of Scottish Hydro-Electric plc. The engineers of Loch Sloy of the 1940s have been replaced by a new breed, differently motivated in the world of advanced state-of-the-art technology and business. The old culture of which I was a fleeting part in 1948, was remarkable for its own endeavours in minimizing the effects of hydro schemes on the environment. The pride of that culture in 'hydro' was, however, limited to the local effects on loch levels, river flow, salmonid fisheries and the scenery. Today, for good or bad, these effects are largely accepted as an established feature of the Highland scene.

Furthermore, in these days of concern about atmospheric pollution by power stations, the engineers wear 'hydro' on their lapel with a new-found pride, for in an otherwise polluting industry, hydroelectricity stands clean. In arid areas of the world it has led to serious problems in siltation, flood plain management and agricultural production, notably following the construction of the Aswan High Dam in Egypt. In temperate countries, however, it creates a favourable impression in the public mind. People are worried by the effects on human health and survival caused by emissions from coal-, gas-, and oil-burning stations, and radioactive waste disposal from thermonuclear stations. Emissions of carbon dioxide and methane contribute to the 'greenhouse effect' and global warming; sulphur dioxide and nitrous oxide contribute to 'acid rain', and the acidification of freshwaters and soils.

The 'green' culture of the 1990s possesses an ethos which was largely unappreciated by the old culture of 50 years ago. It pervades the entire electrical industry, being the concern of sales and marketing people, as well as of engineers and architects. It is an ethos which industry generally is still shy of accepting, as acceptance is costly and may result in a fall of production, or the blunting of competitive edge. However, environmental responsibilities are being more and more enshrined in legislation, and are no longer a voluntary matter. In the past, voluntary donations to good conservation causes by industry were invariably welcome and useful to the conservation movement, giving 'green' publicity to the donor. However, the level of such giving was usually on a local scale, unconnected with the large-scale impacts of industry *per se*.

Industry is still running amock in some places, particularly in the developing countries, with the destruction of forests, the erosion of watersheds, the impoverishment of plains, the disruption of lakes and rivers, the stripping of coasts, and the exhaustion of marine resources. In the developed world, Governments are ratifying conventions on conservation and legislating for improvements in their own back yard. In my time, I have seen transformation of attitude from an introverted exercise focused solely on local and national needs to one of international and global proportions. In UK industries, the base line for environmental audit is no longer indigenous; it is that of the European Community and the United Nations.

The 1990s seem likely to be dominated by concern about the environment. The vacuum in world affairs caused by relaxation of the East–West tension in the 1980s is being filled by the joining of hands across the established geographical and ideological divides. There is a sense of commonality in mankind from peoples whose lives are poles apart, becoming aware of their interdependence in the saga of survival on planet Earth. Yet this commonality is not new; it is probably as old as humanity itself.

Over the years which I spent in international conservation, I was always in search of the grounds for a common understanding of conservation with peoples of different ethnic and cultural roots than myself. I soon came to realize that the common ground is indeed the land itself. That was the commodity which was sacred above all others in the minds of everyone, everywhere. Its possession, use, and care, were of unfailing interest whether in a *barazza* in the Masai Steppe or a Select Committee in Westminster. In fishery societies the sacred commodity is the sea. The discussions had the spontaneity of inner passion and conviction which exposed at its core, the ethnic of life.

Coming full circle, at the time I joined the ranks of ecologists in 1948, Aldo Leopold, one of the great ecophilosophers of this century, died in Wisconsin of a heart attack while helping a neighbour fight a grass fire. He was the author of the land ethic – that man was *a part of* and not *apart from* the land. Had he been a fisherman instead of a forester and farmer, I am sure his ethic would have applied to the sea as well. Indeed it applies to the whole planet. If I had read *A Sand County Almanac and Sketches Here and There* when it was first published in 1949, I am sure that my 'conversion' to conservation would have been easier. Leopold's words have remained imprinted in my memory since I first read them long ago:

> A land ethic changes the role of *Homo sapiens* from conqueror of the land-community to plain member or citizen of it.

Oh, where can we find humility enough for that citizenship? Therein lies the ultimate dilemma!

REFERENCES

Boyd, J.M. (1987) Commercial forests and woods: the nature conservation baseline. *Forestry*, **60**, 113–34.

Cairns, W.J. and Rogers, P.M. (eds) (1981) *Onshore Impacts of Offshore Oil*, Allied Science Publishers, London.

Leopold, A. (1949) *A Sand County Almanac and Sketches Here and There*, Oxford University Press, Oxford.

Lyddon, W.D.C. (1981) Control strategies: physical planning, in *Onshore Impacts of Offshore Oil* (eds W.J. Cairns and P.M. Rogers,), Allied Science Publishers, London.

Nicholson, E.M. (1970) *The Environmental Revoluton. A guide for the New Masters of the Earth*, Hodder and Stoughton, London.

Nicholson, M. (1987) *The New Environmental Age*, Cambridge University Press, Cambridge.

O'Riordan, T. (1983) Putting trust in the countryside, in *The Conservation and Development Programme for the UK*, Kogan Page, London.

Payne, P.L. (1988) *The Hydro*, Aberdeen University Press, Aberdeen.

Sheail, J. (1987) *Seventy-five Years in Ecology: The British Ecological Society*, Blackwell Scientific Publications, Oxford.

Thom, V.M. (1986) *Birds in Scotland*, Calton, Poyser.

Case study: research 11

O.W. Heal

Bill Heal is Director of the Natural Environment Research Council Institute of Terrestrial Ecology (North).

11.1 INTRODUCTION

Research, the systematic investigation to establish facts and principles, is not usually associated with ethics, except in the presumption that researchers are honest. Ethical or moral questions concerning genetic manipulation or nuclear power or the like are regarded as exceptions from this generalization because the consequences of research in such areas can have major long-term effects on mankind. In practice, the decisions made by researchers and research managers repeatedly involve questions of ethics, i.e. 'a professional code of practice which is considered correct'.

Although such a code of practice is unwritten or at least not widely recognized, it is an increasingly important part of decision-making in research. Three aspects of a research code (objectivity, priorities and publication) are illustrated in this chapter through a case study on environmental radioactivity.

1. In most environmental research the primary aim is to establish the facts, or at least the rational principles on which policy decisions can be made by those concerned with legislation, planning or management. The basic requirement is the provision of information on which decisions can be made; objective information divorced from advocacy. But how can we ensure that the research is not prejudiced?
2. Commercial pressures and limited resources require decisions about

Environmental Dilemmas Ethics and decisions
Edited by R.J. Berry
Published in 1992 by Chapman & Hall, London. ISBN 0 412 39800 1

research priorities. What are the criteria to be applied and how can we evaluate the social or moral criteria against those of economics or scientific interest? How do we balance short-term financial gain against longer term scientific understanding?
3. The traditional output of scientific research is publication in the open, refereed literature. Increasingly this mode of publication conflicts with the requirements of contractual confidentiality, with the commercial advantages gained by retaining data, and with the public demand for information sooner rather than later.

The history of research on environmental radioactivity at the Institute of Terrestrial Ecology since its initiation in 1978 serves to illustrate the environmental dilemmas faced by individual researchers and managers. However, as we shall see, the same dilemmas recur in many different areas of environmental research and organizations.

11.2 RESEARCH INITIATION

In 1975 the UK Government held a Public Enquiry into the proposal to instal a nuclear reprocessing plant (THORP) at Sellafield on the Cumbrian coast. The enquiry received evidence on the dispersal of radionuclides released, deliberately or accidentally from smoke stacks and sea outfalls, and on their potential transfer through terrestrial and aquatic ecosystems. The primary concern at the time was on the risk of transfer to humans. The environmental evidence was provided by various organizations within the nuclear industry, by Government Departments and by experts from Universities in the UK and from other countries. One of Lord Justice Parker's conclusions was that the UK lacked an adequate research capability which could provide advice to Government, but which was independent of the nuclear industry. He recommended the development of research within universities and institutes. In response, the Natural Environment Research Council (NERC) expanded its research effort in this area, including the establishment of a small group to study radionuclide behaviour and transfer in terrestrial ecosystems, based at the ITE Merlewood Research Station in Cumbria.

Thus while Lord Parker did not reject the advice from experts within the nuclear industry he highlighted the need to distinguish the provision of objective information by independent researchers from that obtained by 'advocates', whether real or perceived.

11.3 RESEARCH DEVELOPMENT

Initial funding of the Merlewood group was from NERC, with additional support from the Department of the Environment. The research strength of the group lay in the expertise within the Institute on the ecology of plants, animals and soils in the natural environment, and on the transfer and transformation of elements within ecosystems. At first, its radionuclide expertise was negligible, but it soon acquired that through collaboration with other organizations, such as the National Radiological Protection Board and various universities. A series of research contracts with the Department of the Environment focused attention on the routes by which radionuclides in liquid low-level wastes discharged into the Irish Sea might be returned to land and retained within salt-marsh soils and vegetation. Further studies explored mechanisms of accumulation and transfer from soils to vegetation, from vegetation to grazing animals, and from invertebrates to predatory vertebrates, notably birds.

By the mid-1980s the credibility of the group in radioecology was well established within the UK scientific community. Research results were presented in contract reports and were published, without any restriction, in the open literature and in Institute reports. Media interest was limited but, as always, very adept at making a story. For example, a factual report of the transfer inland of minute quantities of radio activity by birds feeding on the Cumbrian coast was printed under the headline 'Bird Droppings Radioactive'. However, the policy momentum which had generated the funds to establish the group was overtaken by other priorities. Contract funding which was important to maintaining the research declined.

11.4 CHERNOBYL ACCIDENT

Then on 26 April 1986 the nuclear power station at Chernobyl exploded. Radioactive fallout was transported by the prevailing weather systems over much of Europe; it passed over the UK a week or so after the accident. In the weeks that followed, the Director and senior scientists at the ITE Station were faced with three decisions of increasing complexity.

First, what research, if any, should be done by the Institute on the possible fallout from the Chernobyl cloud? Was it part of the Institute's responsibilities? How high a priority should be given to it in relation to limited resources and other commitments? Here there was no hesitation. Apart from being a potential hazard to human health, this was a major environmental pollution incident which could have effects on other organisms. It also provided an opportunity to improve understanding of radionuclide transfers in ecosystems by analysing the fate of the pulsed input, and the Institute had the relevant radionuclide and ecological expertise for this task. Within a few days a national field sampling programme based on an existing land classification system was designed and, using resources from its six ITE research stations distributed across the UK, grassland vegetation samples were collected and sent to Merlewood for analysis.

Second, how should the Institute's action be coordinated with that of other research groups and those responsible for monitoring of environmental radioactivity? How best could information be provided to Government to assist in their immediate policy decisions? Line management responsibility was through the Research Council to the Department of Education and Science but the key organizations were the Department of the Environment, the Ministry of Agriculture, Fisheries and Food, and the National Radiological Protection Board. Other organizations with major research capabilities such as the Atomic Energy Authority (Harwell, Aldermaston) and British Nuclear Fuels were also involved. It took some time for clear responsibilities and channels of communication to be established (Wynne, 1991) and for the ITE data, which were beginning to show distinct hot spots, to be assimilated by the relevant authorities.

The problems of official, sometimes conflicting, communication (Wynne, 1991) were exacerbated by a third problem: the rapid and intense pressure from the public and media for information on the levels and distribution of contamination, the risk to human health, and the precautions that should be taken. The pattern of contamination emerged slowly, given the inevitable time lag in planning, collecting and analysing samples. How and when should information be released to the public given the incompleteness of the data, difficulties in predicting the mobility of the radionuclides and uncertainty about the risk to human health?

The immediate question of contamination of milk by ^{131}Iodine was not critical because of its short physical half-life of 8 days and the limited deposition that occurred on agricultural land used for milk production. Direct monitoring of milk supplies was undertaken by MAFF and NRPB. The main concern was with the longer-lived isotopes, ^{134}Caesium and ^{137}Caesium. It was decided that: (1) the existence of an urgent sampling programme should be public knowledge, but (2) no data would be made public until there was a clear picture, and (3) no comment would be made on health hazards or potential precautions, questions on which would be referred to the Government spokesman.

Open communication through Government channels was maintained but the lack of generally available information meant that there was considerable ill-informed comment in the media. Government restrictions on sheep movement and sales were in place by June but there was uncertainty about the length of the ban and little information available on the longer-term environmental behaviour of radiocaesium in the semi-natural upland ecosystems that had received the highest deposition levels.

By June the Director of ITE, John Jeffers, decided that controlled publication of results was justified and a factual article was published in *The Guardian* on 25 July, including the first map of ^{137}Caesium concentrations in vegetation in the UK (see Fig. 4.3, p. 64). Publication resulted in 'some discussion' in official circles and there was a brief outburst of interest from the public, but this soon faded.

The general picture of contamination from Chernobyl was established quite early but the details of local distribution were complicated by considerable variation over distances of a metre or so resulting from small-scale differences in vegetation, soil and water movement (Howard *et al.*, 1987). However, the key issue was how long would radiocaesium continue to be transferred to sheep. Results from bomb test fallout studies in the 1960s suggested that radiocaesium would soon be removed from vegetation surfaces and immobilized by adsorption onto clay minerals in the soil. This prediction was supported by experience of fertilizer retention in lowland agriculture. However, it soon became apparent that upland vegetation retained radiocaesium much longer than expected, and the low clay content of much of the upland soils was retarding immobilization. The need for further research was clear, but what research and who was going to fund it? (Howard and Livens, 1987.)

11.5 POST-CHERNOBYL RESEARCH

By late 1986 four types of research were being recognized.

1. Monitoring of the progress of various radionuclides from the Chernobyl fallout in the environment and particularly in products affecting man. This was undertaken by NRPB and MAFF as part of their statutory responsibilities;
2. Development of countermeasures to prevent the radiocaesium passing into humans e.g. transferring sheep to uncontaminated pastures (Howard *et al.*, 1987) or preventing uptake of radiocaesium by applying binding agents such as bentonite to contaminated vegetation (Beresford *et al.*, 1989);
3. Analysis of the mechanisms and pathways of radiocaesium behaviour in the environment to improve targeting of ameliorative practices and policies;
4. Use of the opportunity of the Chernobyl 'pulse' to improve understanding of the dynamics of radiocaesium in different ecosystems.

The four types were a sequence from applied to strategic research, with funding responsibilities moving from the statutory responsibilities of MAFF and other Government Departments to the 'grey' area of strategic support including Research Council funding. Within ITE new projects were initiated mainly through contracts with MAFF, redirection of research supported by the DOE and later by new studies for the Scottish Office. NERC also funded a 'Special Topic' programme which used the pulse of radioactivity to stimulate more exploratory research; ITE gained a number of postgraduate studentships and post-doctoral fellowships jointly with universities and other institutes.

But as the options and urgency for further research on Chernobyl decreased, UK funding also declined. However, EC contracts have resulted in comparative studies across Europe, a position stimulated by glasnost which has allowed examination of highly contaminated situations near Chernobyl itself. Research organizations such as ITE are now faced with decisions on the level of research on environmental radioactivity that can be sustained in relation to competing priorities such as land use, climate change, and other forms of pollution. What size of team should be employed, given the small funds available; does the team have a critical mass, allowing flexibility to expand when the need arises? Many lessons were learned from Chernobyl but it was a specific incident with a limited suite of radionuclides, deposited in particular environments at a particular time of year. One thing is certain: the next incident will pose different questions. This means that there is a strong case for maintaining an environmental radioactivity facility especially as there are few organizations which have the relevant expertise in the subject.

There is also a strong case, highlighted by sociological research after Chernobyl, for much more open communication with the public. An analysis of farmers' experience showed that scientists not only underestimated the ability of lay people to judge risk and uncertainty, but that farmers' informal knowledge 'needed to be integrated with more abstract and formal scientific knowledge to create an effective response' (Wynne, 1991).

11.6 GENERAL CONSIDERATIONS

Apart from NERC support, funding has hitherto come mainly from government and EC sources, i.e. outside the industry. Should researchers increasingly look towards NIREX, BNFL, Nuclear Electric and other parts of the nuclear industry for support? The principle of the polluter pays is applicable but the original rationale for establishing the research group was to provide expertise independent of the industry. Can a research group retain its objectivity when it is under contract, whatever the subject? The researcher would answer yes, but commercial and other pressures can be considerable and the public is unconvinced.

The pressure to obtain contract funding for research has increased considerably in the last decade in the UK and in many other countries. A natural response to this pressure is to emphasize the importance of one's own subject in order to gain funds. Where is the boundary for individuals and organizations, between reasonable publicity and unethical hype?

A further question concerns open publication of results when research is done under contract. This should be a straightforward issue to be settled by negotiation with the researcher free to publish the research separate from the value judgement, interpretation or advocacy of the contractor; this is a simple guideline, but often difficult to negotiate.

The questions of objectivity, publicity and publication apply to most contracted research, whichever organization is commissioning the work and on whatever subject. The integrity of the researcher is essential, and the design of experiments and interpretation of results must be unbiased and logical. And to retain integrity the researcher also has to eschew advocacy of a particular policy or practice.

The question of the objectivity of the researcher is becoming increasingly important as environmental issues are more widely debated. A distinction has to be made between three interacting groups of people (Fig. 11.1), all of whom may be scientists but have distinct roles as explored in Lord Ashby's Preface (p. x).

The objectivity and integrity of the researcher is a basic requirement with peer review of both research proposals and publications providing a key mechanism for quality assurance. It is a time-honoured part of the practice of science but is itself open to misuse through downgrading a competitor's work or making use of the proposals of others.

In summary, research within ITE on environmental radioactivity highlights some of the general issues facing researchers and research managers: independence, objectivity, priorities, communication and publication of results. Within the context of a Code of Environmental Practice (pp. 253–60) these issues are clearly covered by the principles of open and honest debate (4), of obligation to review standards and practices (5), and of acceptance of responsibility to collect and disseminate information (7). These principles are readily translated into practice particularly through open publication and use of peer review. Decisions on the relative priorities of research topics, particularly when the problems are long term, are probably the most difficult. Various criteria can be applied: the extent to which the subject is well researched, the

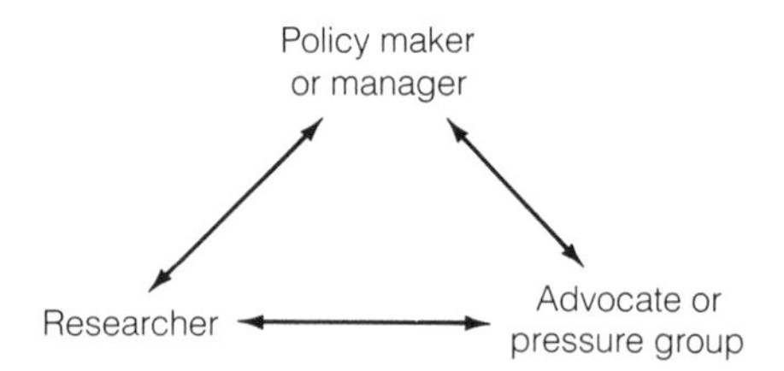

Figure 11.1 Three interacting groups of people.

opportunity for predictive research, the opportunity to identify fundamental mechanisms, the degree of application of results which can be expected, and the strength of the research group. Much, though, depends on personal judgement, critically based on a principle of open consultation.

ACKNOWLEDGEMENTS

I am very grateful to S.E. Allen, R.J. Berry, A.D. Horrill, B.J. Howard, J.N.R. Jeffers and P.B Tinker for their constructive comments and factual information.

REFERENCES

Beresford, N.A., Lamb, C.S., Mayes, R.W. *et al.* (1989) The effect of treating pastures with bentonite on the transfer of ^{137}Cs from grazed herbage to sheep. *J. Environ. Radioactivity*, **9**, 251–64.

Howard, B.J. and Livens, F. (1987) May sheep safely graze. *New Scientist*, No. 1557, 23 April, **114**, 46–9.

Howard, B.J., Beresford, N.A., Burrow, L. *et al.* (1987) A comparison of ^{137}Caesium and ^{134}Caesium activity in sheep remaining on upland areas contaminated by Chernobyl fallout with those removed to less active lowland pasture. *J. Soc. Radiol. Prot.*, **7**, 71–3.

Wynne, B. (1991) After Chernobyl: science made too simple. *New Scientist*, No. 1753, 26 Jan. **133**, 44–6.

Case study: economics – the challenge of integrated pollution control

12

R.K. Turner and J.C. Powell

Kerry Turner is Executive Director of the Centre for Social and Economic Research on the Global Environment, University of East Anglia, Norwich and University College London. Previously he was Senior Lecturer in Environmental Economics at the University of East Anglia, Norwich. He has been an economic consultant to a number of Government Departments and to national and international agencies such as the National Audit Office, the National Rivers Authority (NRA), the OECD's Environmental Directorate, and the European Commission. He is the author of Sustainable Environmental Management *(Belhaven, 1988),* Economics of Natural Resources and the Environment *(with D. Pearce, Harvester Wheatsheaf, 1990) and a contributor to* Blueprint 2: Greening the World's Economy *(Earthscan, 1991).*

Jane Powell is a Senior Research Associate in the Environmental Risk Assessment Unit at the University of East Anglia, Norwich, concerned particularly with the potential uses and disposal problems of waste.

Much of the analysis in this chapter is drawn from a research project co-funded by Her Majesty's Inspectorate of Pollution and the Economic and Social Science Research Council.

Environmental Dilemmas   Ethics and decisions
Edited by R.J. Berry
Published in 1992 by Chapman & Hall, London. ISBN 0 412 39800 1

12.1 INTRODUCTION

Two important UK Government environmental initiatives have come to fruition in recent years: in September 1990 a major White Paper on the environment (*This Common Inheritance*) was published and in November 1990 the Environment Protection Bill went on to the statute book. From industry's perspective the most important issue in the latter was the new system of Integrated Pollution Control (IPC) (as set out in DOE, IPC-A Practical Guide, DOE 1991). IPC is meant to be a multimedia approach to pollution control, such that in cases where substances are released to more than one environmental medium, the Best Practicable Environmental Option should be selected by the polluter. It still, however, retains the conventional technology-based approach to pollution control, based on the core concept of Best Available Technology Not Entailing Excessive Cost (BATNEEC).

At the international level, the European Commission (EC) has implemented an Environmental Impact Assessment Directive (the regulations came into force in the UK in 1988) and is considering a series of possible Directives on aspects of waste management, environmental auditing and integrated permitting. In addition, the EC has begun issuing 'Euro-BATNEEC' notes for certain processes and substances. Both the UN Economic Commission for Europe and the UN Environment Programme have continued to promote cleaner technology initiatives which have been linked to more general trends towards product life-cycle analysis ('cradle-to-grave'), ecolabelling and environmental auditing.

Overall, both the style and content of environmental protection policy is slowly being reorientated in many developed countries. In the UK, IPC is meant to enable the control agency (Her Majesty's Inspectorate of Pollution, HMIP) to move beyond pollution control in specific processes and plants; environmental damage impacts around a plant must now also be taken into account. Meanwhile, a 'precautionary' approach to pollution control, combined with 'a duty of care' in matters of waste management has gained ground. Pollution control authorities are seeking to develop more integrated approaches (i.e. spatial integration across environmental media; administrative/institutional integration; and data integration). There has also been a growing interest in environmental auditing and the adoption of a 'cradle-to-grave' production/product assessment perspective. The development of managerial accounting systems to assess the economic and social costs and benefits of clean technologies can be seen as a positive factor in the stimulation of such technologies (especially process changes as opposed to end-of-pipe measures).

Nevertheless, there are limits to this integration. It is also the case that the precise nature of the accounting system that will be required for a more comprehensive environmental assessment has yet to be properly understood and appreciated by policymakers.

12.2 MEANING, RATIONALE AND LIMITS OF THE TERM 'INTEGRATION'

During the 1970s and 1980s the goals of environmental policy in industrialized economies have been slowly evolving. Most environmental decision-making has proceeded piecemeal, i.e. by way of compartmentalized and only loosely coordinated effort directed at specific issues and problems. Nevertheless, the general objective has been an 'improved' environmental planning and management process, with greater emphasis on long-term considerations. 'Improvement' is taken to include goals such as better ambient environmental quality states; reduced risks to human health; and a more efficient, i.e. cost-effective, pollution control system.

For some, the 'improvement' route leads eventually to an ideal integrated comprehensive decision-making process, in which problems are considered in their interrelated, interconnected totality (Bartlett, 1990). Such a comprehensive process, even if it were possible, may well be both impracticable and undesirable. In the USA, the 'cradle-to-grave' philosophy of waste management has imposed massive reporting requirements and expensive controls, but achievements have been limited (Council on Economic Priorities, 1986). On the other hand, the problems of the overly-restrictive past approach are becoming obvious and in many cases likely to become more severe in the future.

The regulatory systems that have been developed so far are beset by 'failures' (Turner, 1991a). The public does not seem to believe that current institutions protect their environment and their health, particularly where toxic substances are released. While past regulatory policy is perceived as moderately expensive, without reform it is likely to get more expensive and less effective. Its inflexibility may also have impeded technical innovation and threatened industrial competitiveness. The primary goal of environmental policy in the 1990s should be a pragmatic balance between piecemeal, expedient and fragmented decision-making and fully integrated comprehensive decision-making. More attention and resources will have to be devoted to understanding the potentialities, limitations and institutional requirements of more integrated and inclusive control systems.

The future control of micropollutants (many of them toxic substances) will be more difficult and more expensive than past pollution control efforts have been. The siting of treatment facilities (e.g. co-disposal landfill sites and high-temperature incinerators) and other related decisions will be paralysed without clearer and more workable government guidance (Lave and Malès, 1989; Opschoor and Pearce, 1991; Turner, 1991b). Society must be allowed explicitly to confront the complexities and trade-offs inherent in environmental decision-making;

it must be recognized that pollution-control policy inherently exposes contradictory ethical, political and economic efficiency implications, all of which have to be managed amidst scientific uncertainty (Lave and Malès, 1989).

The notion of balance, the need for a more explicit recognition of trade-offs and uncertainties and the fostering of democratic debate about such issues are the implicit themes running through this chapter.

12.3 POLLUTION CONTROL AND RESIDUALS MANAGEMENT 'FAILURES'

Rational decision-making about pollution control and waste management is beset by a series of 'failures' (Turner and Powell, 1991). These 'failures' are ubiquitous but vary in severity and extent from country to country. Four basic categories of failure can be distinguished:

12.3.1 INFORMATION FAILURE

Databases covering plant mass emissions, pollution transfer, disposal and damage costs are deficient. For example, while there are considerable data on Municipal Solid Waste (MSW) and industrial hazardous waste generation and management, most countries lack a single database that is national, comprehensive and current.

Within plants, a more intensive monitoring programme needs to be established to get accurate baseline data on the constituents of all waste streams. Reliable data are most often available on solid and liquid wastes requiring outside contractor disposal but detailed information on effluents to surface waters and emissions to air are generally calculated rather than measured; adequate data on 'fugitive' emission sources are often absent.

The UK lacks acceptable projections for the quantities and composition of MSW which are essential for the planning of an efficient and economical waste management system. A recent official survey of 100 landfill sites revealed an inadequate statistical record of both inputs and monitoring of site conditions. This basic information failure compounds the difficulties that the public, industry and government face in the vexed and often conflict-ridden context of hazardous waste and its management.

Environmental health risks, of natural or human origin, are an ever-present feature of human life. As economies have industrialized, the nature of these risks has changed from past concerns over infectious diseases, through more recent chemical and radiation exposures in the workplace, to current worries over 'environmental' (non-occupational) toxic exposures. This shift in public health focus has produced a reduction in concern about acute illness from relatively high doses of toxic exposure, towards worries about delayed health effects from low-dose exposure (perhaps for years). Risk assessment (typical of hazardous waste facilities) with respect to the latter situation can be an ambiguous process and will not produce the precise answers that society demands.

Given the nature of our economic system and its technological basis, we cannot expect the material benefits of this system, and simultaneously enjoy zero environmental risk founded on zero exposure to pollution. Some sort of *risk–benefit balancing process* is required in which 'acceptable' trade-offs between risk levels and the costs of reducing exposures are struck. But the IPC regulatory approach is vague and of little help in any definition of '*de minimis*', 'significant' or 'acceptable' risk.

Waste disposal facilities in particular have suffered from the NIMBY (not in my back yard) syndrome, with health risk perception usually at the centre of people's concern. But why is it that people perceive the health risks to be unacceptable when formal analysis does not confirm these perceptions? Experts tend to use the 'relative-risk' approach, i.e. the risk posed by toxic chemical exposure from a waste site versus risks like smoking, alcoholism, poor diet, and traffic accidents. Opinion polls show that chemical waste disposal is at the top of public concerns, but risks from such sites (active and inactive) only rank eight and 13 on the US Environmental Protection Agency's list of 31 cancer risks.

There is, however, a psychosocial basis for the NIMBY syndrome. Waste facilities defined as hazardous are inherently stigmatized and therefore classed as undesirable. Deeper and wider social concerns may also underlie local opposition, such as invasion of homelife and territory, loss of personal control, stress and lifestyle infringement, loss of trust in public agencies and lack of accountability of the 'system'. Here the waste site is merely a catalyst of unlocking concerns about trends in society in general.

The database required to assess the wider 'regional' environmental impacts of residuals discharged/emitted from industry is particularly deficient. There is a lack of information about pollutants being discharged or emitted, the magnitude of pollution loadings, and spatial and temporal variations and trends. The science of ecotoxicology is underdeveloped, with relatively little research being directed so far at damage-impact identification and measurement on the ecosystem scale. Finally, not enough is known about the structural and functional value of ecosystems. The question of the intrinsic value of such natural systems represents a further controversial element in the overall debate (Turner, 1991b).

12.3.2 LACK OF 'SYSTEMS' THINKING

Given the diversity of interests involved in or affected by the waste management process, it is not surprising that the debate is often confused and full of conflicting proposals. By and large local authorities and private reclamation firms support MSW recycling centres which are meant to minimize collection costs. The beverage container industry and voluntary groups favour door-to-door collection to maximize participation and material yields. A number of other interests see merit in centralized treatment and recovery plants. Industry's response has been based on the idea of cooperative MSW recycling ventures bringing together retailers, local government and the public. It is argued that finance for such ventures can be raised via voluntary levies on producer firms, e.g. a disposal charge added to the price of new tyres, with the revenue raised being direct towards an expansion of tyre retreading activities.

The overall systems perspective must be kept firmly in mind. Household-based recycling will lower the calorific value of the remaining MSW and increase its moisture content. Residual disposal is consequently more costly. Mandatory deposits on glass bottles may inhibit overall levels of glass recycling. Too rapid a stimulation of recycling can depress prices and lead to a collapse of market confidence. Markets for reclaimed materials are still underdeveloped and constrained by the grade structure of reclaimed materials.

Industrial waste issues are often even more complex. In the USA, while two-thirds of the projects to reduce hazardous chemical discharges to the air are based on reduction at source or recycling measures, a mass emission systems perspective reveals that two-thirds of these measures also ultimately involve end-of-pipe measures, mainly incineration.

Finally, these management systems and the related economic system are underpinned by ecological systems. An explicit recognition of the ecological foundations is required. In practical terms this means a need to extend the spatial context of pollution control, at least up to a 'regional' level.

12.3.3 INSTITUTIONAL FAILURE

Waste management in the UK has been inconsistent and complex involving District Councils, County Councils and the Central Inspectorate. The division of collection and disposal between different agencies has contributed to poor data. The shared role of licensing and policing of waste sites is a contributory factor in the inconsistency of standards across sites.

The advent of IPC will add a new level of complexity to the decision-making process for major industrial projects. Local authorities may well have to take decisions on planning applications for such projects without a clear indication whether they comply with new pollution control rules. Under the 1988 Environmental Assessment Regulations, HMIP may be asked whether it considers that an environmental statement should be prepared for proposed projects; where an environmental statement is required, it must pass information relevant to the project to the applicant on request. But HMIP is also a statutory consultee in the assessment process. It is likely to be very cautious in commenting on environmental statements so as to protect its own role as regulator under separate legal provisions. Other statutory consultees, such as the conservation agencies, may be equally reticent, wishing to keep their 'powder dry' for possible public inquiries at a later stage.

There is no doubt that Environmental Assessment will have to play a significant role in the achievement of the objectives of IPC for new plant, especially if the BPEO for residuals management is to be identified. But there is no formal requirement or mechanism for Environmental Assessment in the IPC authorization procedure relating to new plant. One of the most potentially innovative features of IPC is that it could enable HMIP, in pursuit of the BPEO, to influence the process technology that is adopted. A bias towards source reduction measures as opposed to end-of-pipe palliatives could therefore be fostered. However, where formal involvement of HMIP occurs after planning permission has been granted (whether following Environmental Assessment or not) and the plant is being built to a preselected design and process technology, then it may be too late for HMIP to exert an influence. Options which might be more 'environmentally effective' will, as a result of institutional failure, no longer be 'practicable'. The case for an integrated, coordinated approach, perhaps with one mandatory Environmental Assessment encompassing both planning and operational pollution control stages for new plants, is a compelling one.

12.3.4 LACK OF AN APPROPRIATE DECISION FRAMEWORK

A number of possible decision frameworks can be distinguished: technology-based standards, risk-based regulations, no (zero) risk provisions via bans or prohibitions, risk–benefit analysis and cost–benefit analysis linked to economic incentive instruments (Lave and Malès, 1989). Striving for *zero risk* as a general strategy can be disregarded; a no-risk criterion is too simplistic for regulating important or popular substances or practices.

Technology-based standards relating to 'Best Available Technology or Technique' (BAT) have been, and continue to be, the favoured approach in all industrialized economies. The problem is that at any given time a best technology does not exist. Another layer of control can always be added, or there is a bench-scale control technology under test that promises more effective pollution control. In the USA, EPA chooses as BAT a technology that is commercially available, reliable, has an 'acceptable' level of control and is available at a 'reasonable' cost.

The UK's IPC will require an assessment of the primary routes by which a given prescribed industrial process is likely to affect the environment (whether globally, regionally or locally). A polluter then has to justify the chosen process/technique in relation to the impact assessment. Where a particular local environmental asset or sensitivity exists, e.g. proximity to a nature reserve or some other designated environmental or amenity site, the operator must demonstrate that the proposed process/technique takes adequate account of it. This is likely to prove very difficult given the paucity of ecotoxicological models and data. Under IPC a technique will be deemed to be 'available' only if it is 'procurable' and also if it has been developed (or proven) at a scale which allows its implementation in the relevant industrial context with the necessary business confidence.

The NEEC ('not entailing excessive cost') portion of BATNEEC means that the presumption in favour of BAT can be modified by two sorts of 'economic' considerations, i.e. whether the costs of applying BAT would be excessive in relation to the environmental protection achieved, and whether they would be excessive in relation to the nature of the industry.

The main drawback of this technology-based standards approach is that it is vague in terms of its guidance on questions of how clean should the environment be made, what are acceptable risks, and what are excessive costs (e.g. what do we do if plant closures might be threatened in areas of high unemployment)? BAT is defined by engineering judgments that weigh human health and other environmental benefits against costs. This type of standard provides an implicit rather than an explicit answer to difficult ethical, political and economic efficiency questions. The inevitable value judgments and social trade-offs between risks and costs are masked and the prospect of more intense nimbyism is encouraged.

Risk–benefit or *Cost–benefit* approaches do at least require explicit balancing of risks, costs and benefits, with the necessity to make trade-offs fully recognized. They can also play a valuable heuristic role in the decision-making process. The acceptability of a regulatory standard is a complex value judgment involving the risk level, the benefits of the regulation, and the cost of further risk reduction. As Lave and Malès (1989) have put it: 'automobiles have a high enough benefit to be worth the risk but if they could be made safer at trivial costs, then the current risk level would not be acceptable'.

Public policy concerning pollution control and waste management (including hazardous waste) can be judged from five different viewpoints: economic efficiency, distributional equity, administrative simplicity, institutional (agency and public) acceptability, and risk reduction. A multicriteria approach illustrates the challenge of designing programmes to meet more than one (often conflicting) policy objective (Table 12.1). Clearly no approach is rated highly on all five criteria. In the past, a

Table 12.1 Comparative Evaluation of Different Decision Frameworks*

Regulatory Approach	Economic Efficiency	Equity	Administrative Simplicity	Acceptability	Risk Reduction
No risk (bans)	very low	very high	high	very high	very high
Risk-based (regulations)	low	high	high	high	high
Technology-based (standards)	very low	low	very high	high	high
Risk–benefit analysis	high	low	low	low	low
Cost–benefit analysis (augmented by economic incentives)	very high	low	low	low	low

* Adapted from Lave and Malès (1989).

combination of no-risk and risk-based approaches (directed at selected toxic substances problems) and technology-based standards has been the preferred strategy. But as society has demanded increasingly stringent controls, costs and regulatory intrusiveness have become much more apparent.

We believe there is a pressing need to move towards cost–benefit and risk–benefit frameworks, which make explicit the balancing process required and the social trade-offs that need to be faced. For specific toxic waste situations, a risk-based approach could be adopted with a governmentally-defined safety goal, or at least a process for arriving at the specified goal. Pollution control and waste management issues have a technical-scientific foundation but they also involve society's values about health, standards of living, and ambient environmental quality. Too little attention has so far been given to public participation in the control strategy, and social costs and trade-offs have been implicit rather than explicit. Even local authorities and their various agencies lack a proper information base to assess risks and trade-offs. There is a particular lack of awareness about toxicological and ecotoxicological impacts and standards at this level in the regulatory system in the UK.

12.4 FINANCIAL VERSUS ECONOMIC VIABILITY

The adoption of a risk–benefit/cost–benefit decision framework would help to clarify a persistent misunderstanding about the distinction between financial profitability and economic efficiency that has been prevalent in regulatory circles. Under the BATNEEC and BPEO-based approach the terms 'practicable' and 'excessive cost' can and have been interpreted (with no great consistency) as financial concepts (with regard to the costs of pollution control measures to polluters and/or in the context of a rationale for the adoption of cleaner technologies); or as economic concepts, in which wider social costs (e.g. impact of pollution damage) are included.

The current UK interpretation of NEEC under IPC is hinted at in the following quote from the DOE (1988) *Integrated Pollution Control: A Consultation Paper*:

> HMIP would examine a plant operating a scheduled process, employing BATNEEC to prevent emission/discharge of pollutants. If significant air pollution still resulted then further controls would have to be applied.

NEEC is here based on financial rather than economic costs; the prime consideration is the financial cost of pollution abatement to the firm concerned and the consequent impact on that firm's relative competitive position in its UK market sector. A further consideration may be the firm's competitive position *vis-à-vis* its EC competitors. Just how one then decides that the residual pollution (after the imposition of BATNEEC) is or is not 'significant' is not spelt out. The EC now seems to believe that the difficulties associated with the concept of NEEC are such that future policy should be established around BAT only.

A risk–benefit/cost–benefit approach could play an important role in rigorously redefining NEEC in social costs and benefits terms. Thus waste disposal authorities (WDAs) are often required to prove financial profitability, i.e. that costs must be outweighed by the benefits/revenues of any management scheme. But the requirement *should be* that the introduction of such a scheme into the management system would reduce overall net social costs (i.e. private plus external costs) of the system. A recycling scheme, for example, should be compared with the currently available least-cost disposal alternative (landfill in the UK). A recycling scheme will generate overall net social benefits, as long as the net social costs of the least-cost disposal option are greater than the net social costs associated with the combined recycling and disposal option.

12.5 MARKET FAILURE

The disposal option has not hitherto been fully costed in social terms because of a failure in the market process itself. In general, individual functions in the waste disposal service are not correctly priced. Typically, the waste collection and disposal service for MSW is paid for via general taxation. Thus waste items do not carry a price tag for the individual waste generator. Even industrial waste generators using contractor services do not typically pay the full social costs of waste disposal. The result is that the financial costs of waste disposal to waste generators

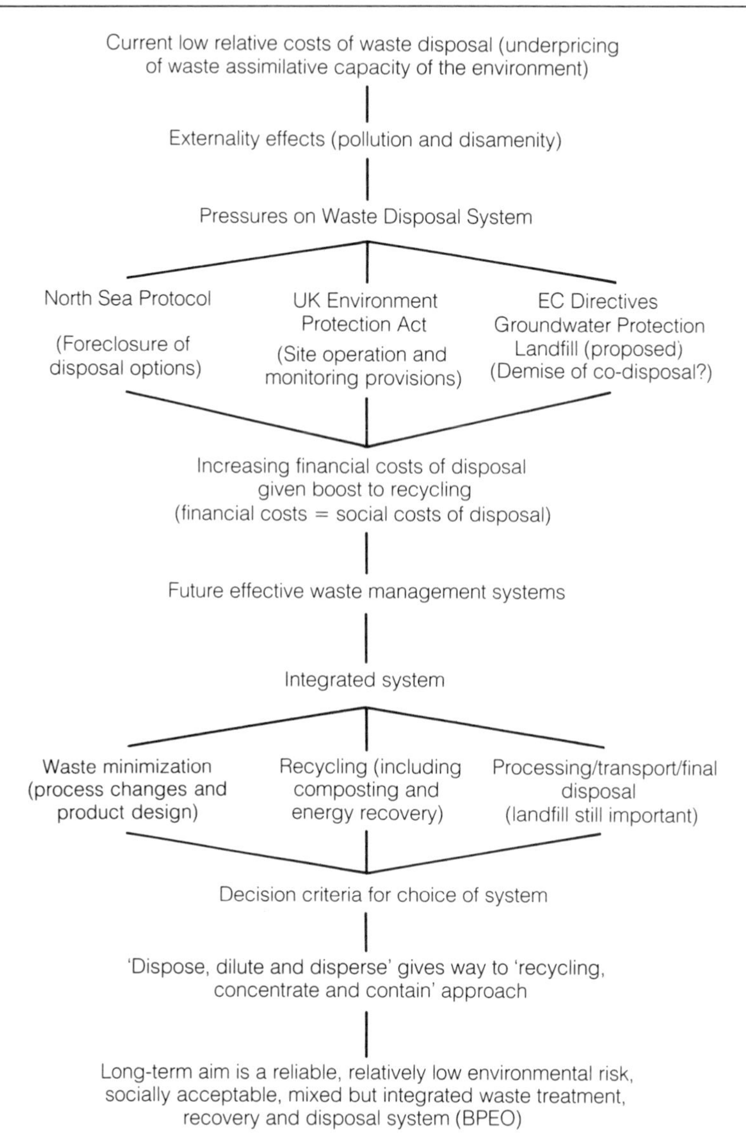

Figure 12.1 Future strategy for solid waste management.

continue to be relatively low, and represent an under-pricing of the waste-assimilating capacity of the environment.

Industrial waste generators have been encouraged to reorientate their residuals management policy away from end-of-pipe control towards the elimination or reduction of waste at source (source reduction). One of the principal tenets of this *'cleaner technology'* movement is that environmental protection and financial profitability goals can be pursued together.

While there are many opportunities for firms to meet such twin objectives, fundamental source reduction measures will often prove to be financially profitable. It may, however, be the case that promoting a more 'environmentally friendly' image will still bring long-term increases in market share and profits for firms adopting 'cleaner technology'. Moreover, as waste disposal costs rise over the long run (Figure 12.1), source reduction measures will prove more and more viable, even profitable.

The recent history of a chemical intermediates plant in the UK illustrates the likely responses of industrial waste generators in the future. Up to 1987, organic wastes were incinerated at sea and other wastes were simply dumped in the marine environment; landfill facilities took the rest of the aqueous waste. Because of the disappearance of its traditional waste disposal outlets, the plant had to turn to process modifications, reducing by more than 33% its volume of liquid waste for disposal off-site, despite an increase in production. The new waste management strategy employed by this waste generator is based on process changes, an on-site incinerator, and an off-site neutralization facility.

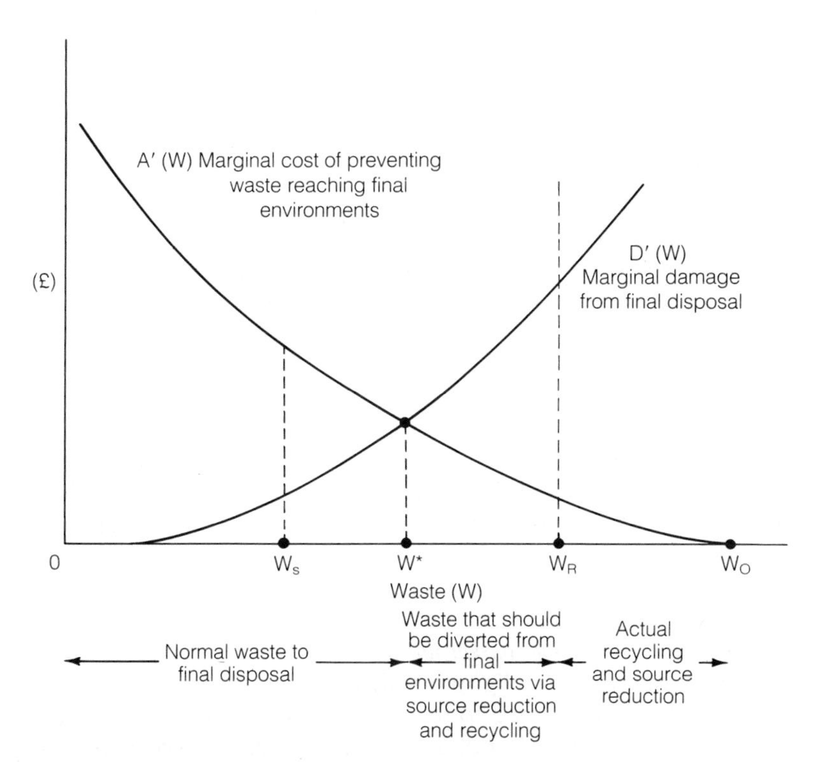

Figure 12.2 Optimal disposal of waste to final environments.

The cost–benefit decision framework can help the general problem of how much waste should be disposed of to its final receiving media, as well as the choice between source reduction and waste recycling options. No method need be chosen unless benefits exceed costs and the optimal scale of activity for any choice is the point where the difference between benefits and costs is maximized. For any given level of waste, the aim is to minimize the costs of managing that waste (Pearce and Turner, 1991).

Figure 12.2 shows the costs and benefits of preventing waste from reaching *final environments* (defined as land, water and air).

12.6 CLEANER TECHNOLOGY INNOVATION AND DIFFUSION: NEW OPPORTUNITIES FOR MARKET-BASED INSTRUMENTS

The clean technology concept began to be taken into industrial practices in the 1970s. In 1976, the USA company 3M introduced a Pollution Prevention Pays (3P) programme for residuals management. Many other companies, particularly chemical conglomerates, followed suit in the 1980s. But the 3P programme emphasized only waste reduction measures which were cost-effective in their own right. Publication in 1987 of the *Brundtland Report* (WCED, 1987) had a catalytic effect on 'environmental management thinking' and cleaner production policy. The Report contained the message that increased environmental protection and continued economic growth were compatible objectives; the impression was given that the scope for the implementation of 'no regret' measures (e.g. energy conservation measures) was virtually limitless. However, the notion of 'sustainability constraints' was underplayed and the necessity to confront trade-offs seemed to be consigned to the medium or long-term future (Pearce *et al.*, 1989; Pearce and Turner, 1990; Pearce, 1991).

At the level of the individual firm and its regulatory environment, the financial versus economic viability confusion was compounded by all this. In the case of the 3M company, for example, the total waste generated across all its production units in 1989 was actually higher than in 1975, owing to increased total output and the continued availability of underpriced waste disposal outlets. The 'no regret' 3P programme was clearly only the first stage in a waste reduction strategy. Later stages require significant capital expenditure and long pay-off intervals.

The process of innovation in and diffusion of clean technologies has been inhibited by a *double market failure*. Economic theory suggests that where output markets are oligopolized, with relatively few firms undertaking R and D, this can lead to too much (in efficiency terms) R and D expenditure. On the other hand, environmental damage costs have been far from fully internalized in firms' existing technology 'selection environment' (Nelson and Winter, 1977). This implies that firms will do too little

R and D into process and product modifications and will limit their efforts to end-of-pipe measures sufficient to meet prevailing regulatory standards.

Innovations in cleaner technologies have been faced by a range of constraints. From a firm's perspective, 'drop-in' innovations are most easily adopted, as by definition they can easily be embedded in existing production processes. But only end-of-pipe pollution control measures fit neatly into this 'drop-in' category of innovation. Other constraints include relatively small market demand, lack of information about cleaner technologies, and a reliance on institutional and organization change. Government intervention will therefore be required to foster the emergent clean technology movement and correct for market failure.

The IPC policy as currently formulated (i.e. on a technology-based standards approach) lacks an appropriate decision framework and is not sufficient to push polluters much beyond end-of-pipe measures. We have suggested a role for a cost–benefit approach, but this approach will have to be buttressed by a range of *enabling policy instruments*. As a rule, technology forcing standards and pollution charges (or other market-based instruments (MBIs)) will be required, supplemented by better information provision (ecolabelling, environmental audits, etc. (Kemp *et al.*, 1991)). Experience with MBIs to date indicates that pollution charges and taxes have generally been set too low to provide a significant incentive for an extensive shift to cleaner production (Opschoor and Vos, 1989).

Environmental policy can affect the process of technical change during the rule-making period, both through the policy instruments actually used and through enforcement procedures. Regulations oblige industry to take technical action to reduce their emissions to some predetermined 'acceptable' level. This action may take several specific forms, all of them within a definition of 'clean' or 'cleaner technology':

1. Installation of add-on, end-of-pipe abatement technologies that require no substantive change in established production process;
2. Internal alterations in plant activities enabling waste (secondary) materials/energy to be recovered and recycled or reused. These new arrangements do not involve changes in the basic production process (they are known as 'clean techniques');
3. Major changes in production processes ('clean processes') leading to the establishment of integrated low-emission production technologies, designed to minimize waste and pollution at the source;
4. Redesign and substitution of products.

These different pollution control techniques are not mutually exclusive. Thus a clean technology may need an end-of-pipe measure to purify residual pollution. But the introduction of, for example, closed cycle water circuits in paper and board mills, can still involve latent costs. Thus the technically simple, and low input, process of installing new tanks, piping and plumbing facilities necessary for a closed water circuit system can, after a time lag, cause end-product quality deterioration, increased corrosion and slime build-up and odour problems. These costs have to be set against the benefits of reduced water and energy inputs and retention of raw materials previously lost in effluent discharges.

Sometimes it is more effective to install end-of-pipe purification because a more fundamental change in the production process is either too risky or impossible. Thus the promotion of clean technology should be viewed as an extensive programme of environmental technology enhancement (including all measures (1) to (4) above) and should not necessarily neglect innovation in end-of-pipe measures.

Nevertheless, the distinction between the economic short-run and long-run periods is important in this context. The Dutch government, for example, has implemented a combined effluent standards and subsidy programme targeted at the metal plating industry. Over the short run this strategy has led to an 80–85% reduction in effluent emissions, without any apparently significant adverse financial effects on the industry. However, these results have only been achieved at the cost of a significant, long-term, toxic sludge generation problem. Unless the sludge is disposed of (say in controlled landfill with long-term containment measures), there is the danger that cross-media pollution impacts will be exacerbated. The sludge problem could have been alleviated if process changes rather than end-of-pipe measures had been implemented (e.g. membrane filtration and ion exchange units). Over the long run, such clean processes may well have proved more cost-effective.

In general, it would seem to be the case that in the long-run, end-of-pipe abatement measures rather than new clean production processes, result in higher costs, lower profits and consequently less resources for innovation. Clean processes usually involve improvements in input–output and/or waste to output ratios which can lower the variable costs of production significantly (Pearce and Turner, 1984).

Available data in the 1980s indicated that the use of clean technologies is biased towards end-of-pipe abatement measures. More fundamental clean processes only represented some 20% of total pollution control investments.

The more widespread adoption of end-of-pipe technologies has been the direct result of the dominance of the prevailing type of environmental policy. The promotion of cleaner technologies was viewed as a component or derivative of general environmental protection policy (dominated by a regulatory standards approach), not as a specific technology policy in its own right. Governments have tended to focus on short-term environmental regulation, to mitigate the most acute environmental problems (including human health damage). This has led to a reactive (rather than a proactive), effect-orientated policy. In order to meet standards it was only necessary for industry to install end-of-pipe technologies. Thus only a small amount of the available funds were spent on process-integrated, preventative technologies.

Viewed from a firm's perspective there is less risk associated with end-of-pipe measures. They do not necessarily halt production if they break down, although legal penalties may be incurred. When they fail, there is an after sales and maintenance service available from specialist pollution control technology firms. Process-integrated measures are more difficult for these specialist firms to market because they often have plant/firm specific configurations. End-of-pipe measures also do not carry 'perverse incentive' penalties in the form of a 'ratchet effect' imposed from above by the environmental regulation authorities.

Clean technology (clean processes) is most likely to be set up in new facilities and in expanding sectors of economic activity. Small and medium-sized firms have invariably chosen end-of-pipe rather than process-change solutions to effluent emission problems (Parfitt, 1990).

Given that environmental protection is a special constraint on industry, an important question is how and to what extent government regulation and the way it is enforced effects technical change. The Brundtland Commission (WCED, 1987) argued strongly for the promotion of cleaner technologies via the deployment of a much more extensive range of economic policy instruments, rather than relying too heavily on a standards-based approach.

In principle, a variety of economic instruments could be deployed to stimulate the adoption of clean technologies: measures such as loans, subsidies, tax relief and exemptions, grants; or effluent taxes, fines, charges, performance bonds, deposit-refunds; or pollution rights trading systems, bubbles and environmental risk insurance programmes.

Research by the Organization for European Economic Co-operation (OECD) (1985, 1987) supports the general argument in favour of economic instruments, in particular effluent charges, with their continuous in-built incentive for funding more efficient technologies. It is clear that the key feature of any future regulatory strategy must be greater 'technical flexibility'. The less regulations concern themselves with 'prescribed technology' (i.e. 'best available', 'best practicable', 'best available not entailing excessive cost'), the more they will facilitate technical change. As far as environmental standards are concerned, a performance (emission) standard is usually more effective than a process standard.

Flexibility to back up the search for new technical fixes can be introduced in a number of different ways. Compliance time for some emission standards could be extended; or a timetable for effluent charge alterations could be set, covering a period well into the future to avoid the danger of severe 'ratchet effect' mentality in industry. The 'bubble' concept first developed in the USA could be extended. The funds raised by effluent charges could be used to stimulate technical change via subsidies or grants. But a fine balance will then have to be found between the imposition of prescribed technologies by the control authorities and the freedom of innovation that the 'aid recipient' is allowed.

Until recently, only limited use has been made of economic instruments in developed countries. In the UK, Italy and France economic instruments play only a marginal role in environmental policy (e.g. use of water pollution charge revenues in France to subsidize abatement measures). In the USA, financial aid programmes have been even more limited, although 'bubble' and 'offsets' policy has been implemented in selected contexts. In Germany, the main focus has been on subsidies for pollution abatement, while the Netherlands has introduced both financial aid programmes and a system of charges to promote cleaner technologies (Opschoor and Vos, 1989).

Much as they affect manufacturing processes, clean technologies touch on an important aspect of industrial strategies, and the decision whether to invest in them will depend on many factors and not only on the environmental dimension. Impediments to the adoption by industry of clean technologies include:

1. Structural constraints, e.g. the need to amortize equipment already installed in-line with a firm's existing innovation and investment plans;
2. Cyclical constraints: the trend of the market and financial situation of the firm;
3. Commercial constraints: possible difficulties in marketing new processes, etc;
4. Limited availability of clean technologies;
5. Poor information dissemination: this is particularly evident in the context of small and medium-sized firms;
6. Other risks and uncertainties posed by clean technologies: production risks, 'political' risks, limited secondary materials markets, price volatility, ill-suited regulatory standards.

Three characteristics of industry have a decisive impact on technical change:

1. The firm's innovative capacity and innovation strategy;
2. The firm's economic position and whether its sector is expanding, stagnant or declining;
3. The size of the firm and the structure of the industry of which it is a component, e.g. the more concentrated the industry the easier is its access to information on clean technologies; sectors fragmented into small and even medium-sized units (e.g. the metal plating industry)

tend to remain poorly informed about available technologies and have not proved fertile ground for clean technology adoptions.

Government measures to encourage more extensive clean technology adoption will therefore have to include:

1. Financial assistance (maintaining a fine balance between prescription of technologies and aid recipient flexibility);
2. Research and development programmes;
3. Improved dissemination of technical information, either through direct state intervention, or through specialist trade organizations;
4. Adaptation of environmental regulation to favour clean technologies, e.g. more flexible environmental standards, use of economic incentive instruments (charges, levies, deposit-refunds etc.) and fostering of 'bubble' and 'offset' policies;
5. Establishment of specialist bodies or agencies to promote the adoption and diffusion of clean technologies.

The potential importance of economic instruments has now been recognized. It is now necessary to identify more precisely those economic instruments that appear to be most effective in stimulating clean technology innovation and diffusion, and to develop criteria for designing effective economic instruments.

The 1990s will see the further development of business strategies which may offer new points of leverage for environmental protection authorities. Many of the larger corporations and large-volume retailer firms have been establishing networks of supplier firms. Within the network the tendency has been to transfer the quality care of products partly to suppliers, and this could in future include environmental aspects. The networks encompass large and medium to small firms so the integration of environmental aspects into the general business strategy may have a radiating effect on the whole industrial sector. This 'environmental pressure' imposed on 'supplier firms' by larger enterprises could be important given the problem that the authorities had getting smaller firms to adopt clean technologies.

Such a business strategy is 'risk averse' because of the high costs of developing and applying new technologies, combined with the shortening of the product life cycle. Again the environmental dimension may become relevant as firms become concerned to avoid the risk of a product being banned on environmental grounds (or suffer 'consumer resistance'). Governments can enhance cleaner technology adoption by fostering this increased industrial sensitivity. They can improve the dissemination of information about the environmental quality of products and introduce environmental quality hallmarks for them.

Governments could also place a firm under an obligation to communicate its environmental risks to individuals living and working within or around the vicinity of a plant. The USA experience shows that pressure to develop cleaner technologies increased substantially when such risk communication guidelines were implemented.

Insurance companies are also likely to make stricter demands on firms in the future before they will incorporate environmental risks into their insurance policies.

12.7 STAGES IN THE ENVIRONMENTAL ASSESSMENT/ AUDIT PROCESS

The analysis of the cleaner production movement has highlighted, among other things, the importance of distinguishing the short-run and the long-run effects of innovation, and the details of the particular 'environment' in which any given firm/industry finds itself.

A fundamental distinction that has to be drawn at the start of any assessment process is the one between (1) established industrial works, and (2) proposed new works, particularly those on greenfield sites. The 'regional' ambient environmental conditions need to be assessed *either* in terms of the existing conditions into which a new plant/facility is to be introduced, *or* in terms of existing conditions which have been affected over time by a discharge/emissions impact from an established plant. The latter situation includes plant modification/modernization assessments, or the requirements of an increasingly stringent pollution control policy linked to a changing public perception of 'acceptable' environmental quality.

There are at least *three phases* (probably overlapping) in the development of waste reduction programmes (Loehr, 1985).

12.7.1 INITIAL WASTE REDUCTION PHASE

In this situation a range of low-cost waste reduction opportunities remain unexploited, either by default or ignorance. Current waste management practices can be changed via relatively cheap and simple means, e.g. good housekeeping, separation of waste streams, etc. Policy instruments in this context could be limited, e.g. to better information dissemination; government assisted waste exchanges; and increasing the cost to waste generators for landfill disposal, to a level consistent with the total economic (social) cost of landfill options; or the provision of recycling credits.

12.7.2 SECOND PHASE: TECHNOLOGICAL DEVELOPMENT PHASE

In this phase, the least costly waste reduction methods have been exploited and there is a need for waste generators to review and implement more comprehensive strategies, involving significant capital expenditure. In this phase public policies should be oriented, e.g. to R and D support and MBIs designed to stimulate adoption of process changes and modifications. Such support will have to be increased if the next phase is to be a realizable goal.

12.7.3. THIRD PHASE: TECHNOLOGICAL MATURITY

Waste generators in this situation begin to confront the political, economic and technical limits of waste reduction activities. Technological innovation is required, but also a more sophisticated environmental assessment procedure. Risk assessment and management strategies are needed, to attempt to balance the trade-offs between competing social interests. Waste reduction is not an end in itself, and regulatory standards have to face up to health risks and more general environmental considerations. 'Acceptable' limits of waste reduction will have to be agreed, with the help of risk assessment and cost–benefit decision frameworks.

12.8 DATA INTEGRATION AND ENVIRONMENTAL IMPACT EVALUATION

Ideally under PIC, the task of the pollution control agency would be to assess systematically the potential for harm of substances discharged into any environmental medium. Essentially this is a technical/scientific matter and not merely an exercise designed to increase administrative efficiency.

A number of different data sets will require integration if the technical expertise traditionally displayed by HMIP within the plant context is to be successfully extended to an assessment function encompassing a wider 'regional' context. These data sets include:

1. Ambient environmental quality: an inventory of environmental attributes on a 'regional' scale. These are the baseline conditions for any assessment and will have particular relevance for the scope and level of detail required in any subsequent EIA;
2. Toxicological data related to the substances being discharged, and the ecotoxicological implications;
3. Pollution abatement data, relating to technological and financial cost constraints;
4. An inventory of human activities within the region (other industries, residential areas, leisure uses, etc.);
5. Risk/hazard data relating to potential releases of pollution;
6. Data indicating the potential for cross-media transfers between disposal routes;
7. A combined economic-ecological evaluation system.

With the aid of such a database and a checklist of site evaluation factors, HMIP should then be able to assess the 'environmental sensitivity' of sites that come under their scrutiny, but this formidable network of data sources is not currently available at any level in the pollution control system. The basic 'regional' environmental inventory, relevant for a given plant, will also require supplementation by international toxicological data sources and guidelines. These heterogeneous data sets will need to be interpreted and monitored on a continual basis. The international data output (e.g. from WHO, USEPA, UNEP, etc.) will have to be incorporated into a format consistent with UK practice and principles, as well as with EC Directives. Knowledge of this data set will be an important factor in the process of maintaining HMIP and other local agencies, 'pressure' on waste generators to adapt/modify their facilities in line with 'best practice'.

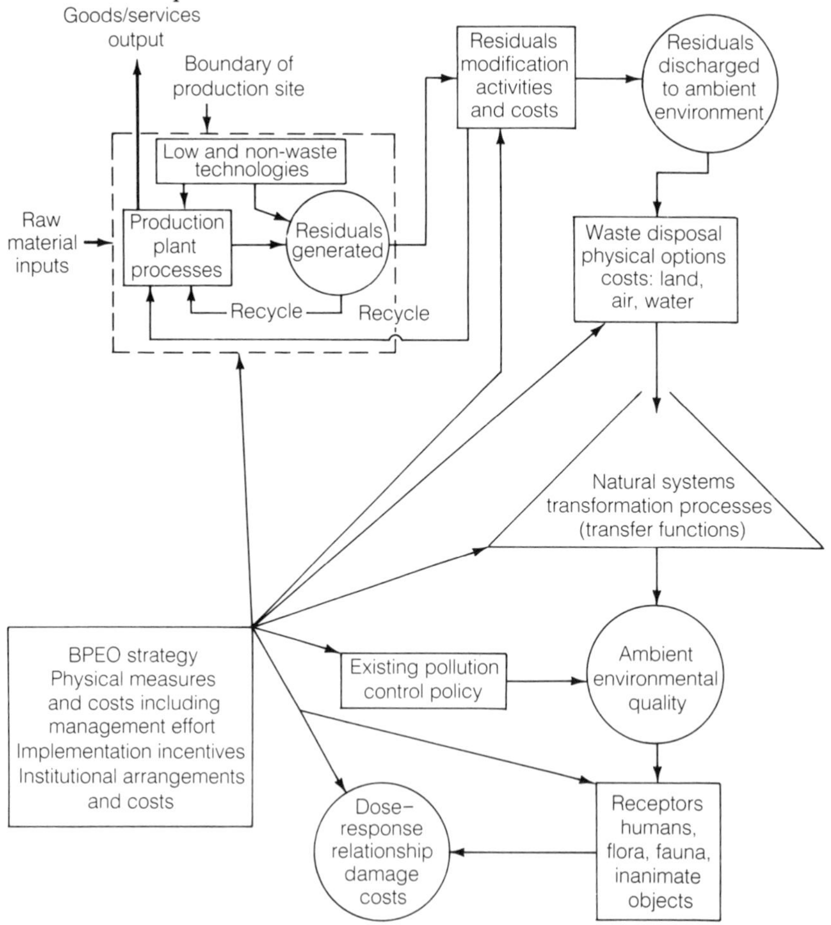

Figure 12.3 Basic BPEO model schema. Reproduced with permission from Pearce and Turner (1990).

A further pollution agency requirement will be access to data covering not only all relevant EC legislation, but also information indicating the likely future directions and scope of this legislation. All these data will be in addition to process data requirements, control cost function estimates and simulated 'macro' economic cost implications.

So-called regional-residuals-environmental quality management models (REQM models) have been developed since the 1970s. These multidisciplinary quantitative models have been designed to analyse industrial activities, the generation and discharge of residuals by activities to the environment, the effects of residuals on ambient environmental quality (AEQ), and the actions that can be taken to manage AEQ at the regional level (James, 1985) (Fig. 12.3).

While a great deal of theoretical work has been done, the practical application of such models has been very limited. Such models are data intensive and expensive. Probably the weakest link in the entire chain of analysis is dose–response modelling and the construction of damage functions. Effects of residuals on human health and on ecosystems are both areas that require much more work. The economic valuation of environmental effects has, however, progressed quite rapidly during the 1980s (Pearce and Markandya, 1989; Turner and Bateman, 1990).

12.9 COMPREHENSIVE ENVIRONMENTAL APPRAISAL

Society has become increasingly aware of damage to the quality of the ambient environment and now demands that waste disposal and/or waste reduction options be both cost-effective and environmentally sound. In principle, wherever it is possible to quantify in monetary terms the impact of a process change or waste discharge/emission on the environment, an economic assessment will be sufficient to indicate environmental acceptability. However, a range of environmental impacts (biophysical and social) may not be amenable to monetary valuation. In these cases, EIA and/or multi–criteria evaluation techniques will be required. Table 12.2 compares the relative merits and limitations of the various decision-aiding techniques.

The term environmental audit has gained prominence in recent years. A given waste reduction and/or disposal system option could initially be deemed efficient or socially worthwhile in terms of its overall demand on natural systems. But this in itself is unlikely to be sufficient: some reductions in the demand for natural resources and environmental services generated by a waste management option involving material/energy recovery, can only be achieved by increasing the demand for non-environmental inputs, e.g. capital. This increased capital demand may well then make its own demand on environmental inputs and also generate wastes.

Table 12.2 Comparison of decision-aiding techniques

Conceptual basis/method	Description	Advantages	Disadvantages	Additional References
1. Standard benefit–cost analysis	Evaluates policies based on a quantification of net benefits (benefits–costs) associated with them	Considered the value (in terms of what individuals will pay) and costs of actions; translates outcomes into commensurate terms; consistent with judging by efficiency implications	No direct consideration of distribution of benefits and costs; significant informational requirements; tends to omit outputs whose effects cannot be quantified; tends to lead to maintenance of status quo; contingent on existing distribution of income and wealth	Pearce and Nash (1981)
2. Extended benefit–cost analysis	Evaluates policies based on a quantification of net benefits (benefits–cash) associated with them	Willingness to pay basis still retained but conditioned by critical natural assets conservation rule; shadow project or offset concept costed into the appraisal process; consistent with judging by efficiency and equity implications	Significant informational requirements; lack of scientific information on natural assets value and substitution possibilities	Pearce *et al.* (1990), Turner (1988)
3. Risk–benefit analysis	Evaluates benefits associated with a policy in comparison with its risks	Framework is left vague for flexibility; intended to permit considerationof all risks, benefits and costs; not an automatic decision rule	Factors considered to be commensurate are not always so; lay and expert perceptions of risk may not be consistent	Fischoff *et al.*
4. Decision analysis	Step-by-step analysis of choices under uncertainty	Allows various objectives to be used; makes choices explicit. Explicit recognition of uncertainty	Objectives not always clear; no clear mechanism for assigning weights	Norton (1984)
5. Environmental impact assessment	Measurement and quantification of diverse environmental impacts	Quantified (non- monetary) data on diverse set of impacts	Diverse data not placed on a common scale; no evaluation possible	
6. Multi-criteria Decision Methods: Lexicographic methods (non-monetary)	Ranking procedure, provides 'best' alternative option on basis of limited number of different critieria	Flexible method, easily adjustable for new options or changes in criteria weights; limited data requirements	Needs a clear exogenous ordering of criteria priorities; equity not considered	McAllister (1980)
Graphical methods (non-monetary)	Illustrates the order of alternatives on the basis of all criteria	Flexible method, results are consistent if trade-off functions (weights) are accepted	Data requirements high. Comprehensive data on ratio/interval scale required for all criteria; weights required for trade-offs; final output masks hidden weights, trade-offs and distributional impact	McAllister (1980)
Consensus-maximizing methods (non-monetary)	Provides 'social weights' for a range of criteria, rather than the optional option	Explicitly incorporates equity considerations	Requires data on individual preference and weights gained via detailed involvement of individuals and groups	McAllister (1980)
Aggregation methods (non-monetary, except for Planning Balance Sheet)	Provides order of alternatives on the basis of all criteria	Flexible methods can consider any number of alternatives; some methods explicitly include equity criterion	Data requirements often high; complexity translates into 'hidden' weights and distribution impacts; subjective judgements given same weight as those based on scientific data	McAllister (1980)
Concordance analysis (non-monetary)	Provides a subset of non-dominant alternatives based on all criteria	Adaptable and methodologically consistent	Complication technique, magnitude of impacts, normalizing fuction and criteria weights required and then 'hidden' in the analysis	McAllister (1980)

Much progress has been made during the last decade or so in the field of economic valuation of environmental assets, with the bulk of the research work seeking to derive 'willingness to pay' monetary measures. Nevertheless, the state of the art in this valuation field is far from being completely satisfactory. While it is in principle possible to categorize all elements in the total economic value (use value + option value + existence value) of environmental resources, it is empirically not possible to measure them all. Non-use values, option and existence values, in particular, are much more difficult to estimate.

Furthermore, there is a need to demonstrate publicly that the best practical environmental option is indeed what it claims to be. Increased public access to pollution control data and to the bureaucratic decision-making process, particularly where large-scale projects are involved, has important implications for EIA methods and decision analysis. Methods are likely to be favoured for their 'transparency' and their ability to provide information understandable to non-experts.

Non-monetary evaluation techniques originated in operations research and developed in response to criticism of monetary methods, particularly cost–benefit analysis. Concerns include the ability of cost–benefit analysis to ascribe partial values to the analysis (Nijkamp, 1987) and the passive role of the decision-maker who is led to his decision (Voogd, 1983).

Since the 1970s 40 or more evaluation techniques have been developed under the umbrella heading of 'multicriteria decision analysis'. These techniques aim to provide a method for the systematic appraisal of alternative projects, plans, etc. against a series of criteria that affect groups or individuals in different ways. Multicriteria evaluation allows the assessment of monetary and non-monetary values in addition to non-numerical evaluations. Most of the differences between the methods arise from the arithmetic procedures used to combine the information from the evaluation matrix (Voogd, 1983).

In the waste disposal decision-making process, multicriteria evaluation allows some insight into the relative importance of financial, resource and environmental considerations (Maimone, 1985). Most of the studies of waste disposal options which use non-monetary evaluation techniques evaluate the location and the options of the projects together (Andreottola, 1989). Maimone (1985) developed a method of planning a solid waste programme in which the first phase is a full evaluation of the waste disposal options independent of their future location and subsequent phases evaluate the site-dependent criteria.

In this study six waste disposal options (incineration reuse-derived fuel (RDF) and landfill, each with (+) or without (−) recycling) were judged against 16 criteria. Cardinal criteria were used where possible but as Maimone (1985) found, the data were often unavailable or insufficiently accurate. In this particular study the 16 criteria were divided into nine cardinal and seven ordinal criteria (Table 12.3). Three of the ordinal criteria could be ascribed monetary values relating to internal and external costs. The remaining 13 criteria were divided into two groups, one relating to resource use and the other to environmental impact.

The allocation of a weight to each criterion is required to define the

Table 12.3 Evaluation matrix used in the multicriteria evaluation of incineration, RDF and landfill of waste with (+) and without (−) recycling. The 16 criteria are listed in column 1

	Incerination+[a]	Incineration−[b]	RDF+[c]	RDF−[d]	Land+[e]	Land−[f]
Cost						
1. Internal cost (1/tonne)[q]	14.07[g]	14.95[g]	7.86	8.23	9.13	9.13
External saving (/tonne)						
2. −transport[h]	0.64	0.59	0.29	0.23	0	0
3. −disposal[i]	2.56	2.41	1.58	1.44	0	
Resources						
4. Land used[j] (ha)	3	3	1	1	13	13
5. % waste eliminated[k]	76	66	53	40	16	0
6. Energy recovered[l] (GJ per capita)	0.56	0.66	0.32	0.40	0	0
7. Percentage materials recovered[m]	20	0	20	0	16	0
8. Waste categories handled[o]	3	3	1	1	3	3
9. Ease of materials recovery[o]	2	2	3	3	1	1
Environmental impact						
10. Transport[n] (km)	0.52	0.66	0.75	0.93	1	0.84
11. Percentage waste incinerated	85	100	18	30	0	0
12. Local air pollution[o]	2	2	2	2	3	3
13. Global air pollution[o]	2	2	2	2	1	1
14. Water/soil pollution[o]	2	2	2	2	1	1
15. Relative concs toxic subs[op]	1	1	1	1	3	3
16. Disamenity[o]	2	2	3	3	1	1

[a] Incineration with electricity recovery, 50% source separation and 80% ferrous metal recovery at plant.
[b] Incineration with electricity recovery and no recycling.
[c] RDF with 50% source separation and 80% ferrous recovery at plant.
[d] RDF with no recycling.
[e] Landfill with 50% source separation.
[f] Landfill with no source separation.
[g] Including flue gas cleaning.
[h] See Section 8.2.2.1.
[i] See Section 8.2.1.
[j] Data for incineration from Coventry plant; RDF data are estimated; landfill data from Department of Environment (1986), and Croft and Campbell (1990).
[k] Including source separation where applicable; RDF data exclude incineration of pellets.
[l] Net energy recovered per capita per year, reduced by in-house energy requirements.
[m] Percentage of total waste stream recovered.
[n] The reduction in the distance waste is transported, and thus the reduction in the pollutants emitted, expressed as a proportion of the distance travelled to dispose of waste to landfill.
[o] Ordinal criteria: values expressed as 1–3, the higher the better.
[p] Relative concentration of toxic substances.
[q] Cost data from Powell (1991).

Table 12.4 Allocation of weights to individual criteria within each of the three groups of criteria

	Weighting
Cost	
1. Internal cost	0.69
2. Transport	0.06
3. Disposal	0.25
Resource use	
4. Land used	0.20
5. Percentage waste eliminated	0.25
6. Energy recovered	0.20
7. Percentage materials recovered	0.10
8. Waste categories handled	0.20
9. Ease of materials recovery	0.05
Environmental impact	
10. Transport	0.16
11. Percentage waste incinerated	0.08
12. Local air pollution	0.16
13. Global air pollution	0.20
14. Water/soil pollution	0.16
15. Relative concentration of toxic substances	0.08
16. Disamenity	0.16

priorities of the decision-makers (Maimone, 1985). However the literature on decision theory reveals that such weights are difficult to determine, particularly within the public sector. There is often no consensus on priorities within a group. Thus a truly representative set of weights is impossible. To overcome this problem, Maimone (1985) created three distinct 'points of view' (national, business-economic, and environmentalist) in which artificially extreme weights were assigned to particular criteria. For the cardinal data, one set of criteria were weighted heavily according to the point of view represented while the remaining weights were distributed more uniformly over the other criteria. For the ordinal sets, priority was given only to the criteria relevant to that point of view, with the other criteria being left out of the evaluation.

In this study the allocation of weights is handled somewhat differently to allow a more comprehensive sensitivity analysis between the three different viewpoints (financial, resource use, environmental impact). First the criteria within each group were allocated weights relative to one another such that the weights within each group summed to unity (Table 12.4). The derivation of these weights was a subjective assessment based on the UK's waste disposal options and regulatory systems.

When only the internal costs are taken into account the relative appraisal scores of the six waste disposal options are:

RDF+	1.0
RDF−	0.948
Landfill+	0.821
Landfill−	0.821
Incineration+	0.124
Incineration−	0

where 1 is the best score.

When the other external costs are taken into account then the position changes with landfill declining considerably relative to RDF (Fig. 12.4; 100% weighting on costs). This results from the savings in transport and disposal costs associated with RDF and incineration. Full details of the internal and external costs of the disposal options can be found in Powell (1991).

The relative appraisal scores for each waste disposal option for each of the three weighting methods are illustrated in Fig. 12.4. For all three weighting scenarios, RDF with recycling remains the best option unless the weight put on resources is very high. However, as the weight given to costs declines the relative merit of incineration increases, while that of landfill and RDF declines. An increase in the weight on the resource criteria relative to costs has a greater impact on the relative position of incineration than does an increase in weight on the environmental impact criteria (Figure 12.4).

Three conclusions are particularly strong. First, RDF with recycling is the best option. Despite considerable variations in the weighting procedure RDF with source separation obtained the highest appraisal score. It was only when more than two-thirds of the weighting was given to resource use that incineration gained the highest appraisal score. Second, recycling in all cases is advantageous to the waste disposal option. This is because it decreases the cost by reducing the mass of waste to be disposed of, increases the recovery of secondary materials and, in the case of plant separated material, provides incineration and RDF plants with an additional income. The third strong conclusion is that by putting increased weight on resource use and environmental impact criteria instead of costs, the superiority of incineration increases and that of RDF and landfill declines. This is because incineration is the best option for maximizing waste volume reduction and inertness, materials separation is a practical option and the environmental impact of incineration can be measured and controlled.

Although RDF is clearly the best option there is an important assumption in the costing of RDF. This is that all the pellets are sold. Most of the RDF plants in the UK have not proved to be viable financially because of limited market opportunities for pellatized fuel. It must also be remembered that while this study attempts to examine the whole system,

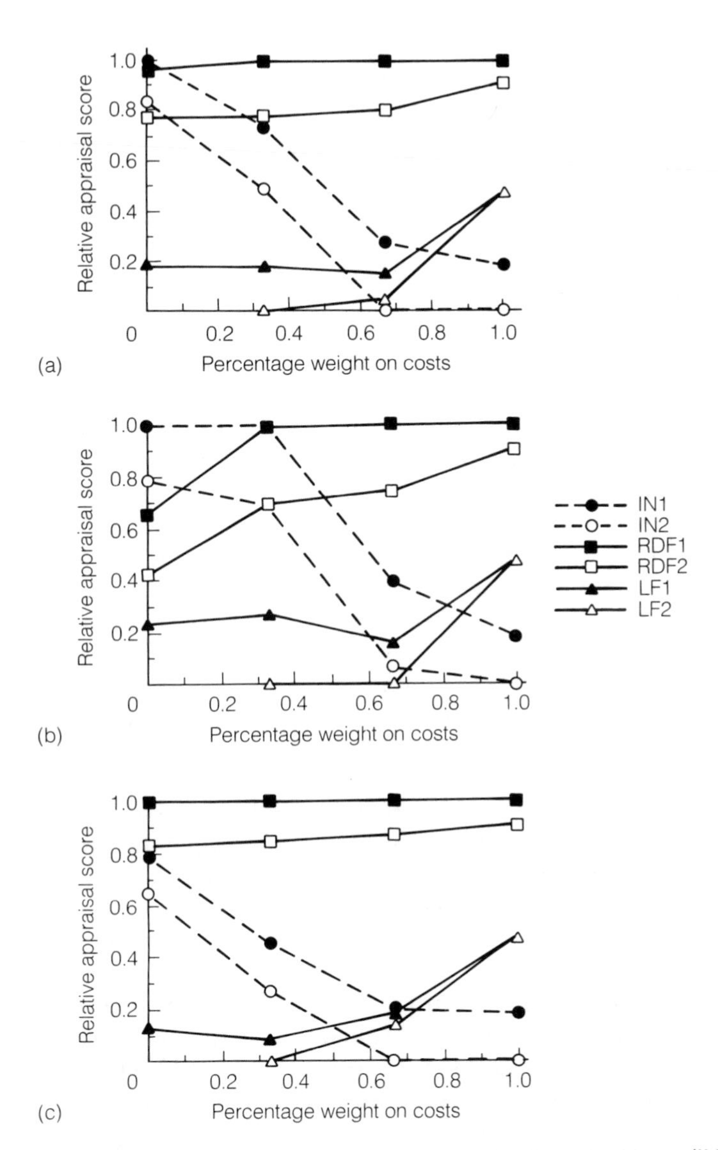

Figure 12.4 The relative appraisal of incineration with (IN1) and without (IN2) recycling; RDF plants with (RDF1) and without (RDF2) recycling; and landfill with (LF1) and without (LF2) recycling; as related to the percentage weight put on costs in the multicriteria evaluation. (a), Equal weighting given to resource and environmental impact criterial; (b), no weight given to the environmental impact criteria; (c), no weight given to the resource use criteria.

from waste generation to the disposal of residue and the production of energy, it does not include the cost of burning the pellets to produce the final energy products as the data are not available. However, in the baseline case the sale price of RDF pellets (£0.63 per GJ) was only a quarter of the price of coal on an energy basis which meant that end users of the pellets will be saving costs rather than increasing them. If this was included in the evaluation it would improve the RDF option.

The multicriteria evaluation technique has a number of clear limitations. There are major uncertainties associated with the method itself, the criteria used, the assessment, and the priority procedure. In addition, there is also uncertainty related to the selection of choice-possibilities or alternatives (Voogd, 1983; Walshe and Daffern, 1990). In the study described, incineration with steam recovery or combined heat and power options were not included in order to limit the number of alternatives. Multicriteria evaluation may give a spurious air of precision to a subject that is still controversial (Walshe and Daffern, 1990).

The derivation of the weights was based on a subjective assessment of the limited data available. However, a more comprehensive assessment could be achieved by means of a survey which might take a variety of forms including paired comparisons, ranking, a multipoint scale (the allocation of scores to criteria) or rating (the distribution of a fixed number of points among a series of criteria). A commonly used method for this is the Delphi Technique, where a panel of experts are used to evaluate the weightings (Sobral *et al.*, 1981). Against this, it has been argued that surveys are too time-consuming and that the most appropriate weighting procedure is to formulate hypothetical alternative qualitative weighting schemes (Voogd, 1988).

Although changes in the weighting in this study did not appear to affect the overall result, different demands on the waste disposal service may well produce different results. For example the production of RDF pellets still requires a commitment to landfill. If this is unavailable or expensive, then the weighting on the criteria concerned with the proportion of waste eliminated would be far greater. Also a bigger population, and thus a larger EFW plant would produce different results; the cost of incineration in particular is very sensitive to economies of scale. Clearly it is difficult to evaluate waste disposal options in isolation. The location of the facility with its inherent conditions and problems can have a major impact on the outcome of an economic evaluation of any type.

However, despite the problems associated with this method of analysis it does provide a flexible structure for comparing options. It is based on principles that are easy to demonstrate, it is relatively cheap, makes explicit the values and norms on which the results are to be based, and is able to illustrate the different outcomes that would occur as a result of modifying the criteria weightings. To some extent multicriteria evaluation can be considered to be a means of structuring a problem rather than finding the solution to the problem (Voogd, 1983). The end results can be considered less important than the learning process gone through to obtain them.

12.10 CONCLUSIONS

Although the notion of an integrated and comprehensive residuals-impact assessment and management system is intuitively appealing, there are conceptual and practical limits to it. The constraints are both practicable (i.e. the data and institutional resource demands) and philosophical (i.e. a fully comprehensive environmental assessment, even if feasible, may not be socially desirable). A balance therefore needs to be struck between a plant-based regulatory pollution control system and a more extensive 'regional' approach to residuals management.

The established residuals management system in the UK is beset by a range of failures: information failure, lack of systems thinking, institutional failure, market failure, and the lack of an appropriate decision framework. Mitigation of these failure problems would lead to a more rational management strategy and also improve its environmental effectiveness.

The conventional technology-based standards approach to regulation tends to shroud social trade-offs, sustainability constraints and scientific and social uncertainties behind a scientific/technical facade. The more widespread use of cost–benefit/risk–benefit decision frameworks would help to make choices and trade-offs more explicit and participatory. Current IPC guidelines offer little in the way of advice in terms of how exactly the system decides what is an 'acceptable risk' or what is an 'excessively costly' pollution control measure: costs and benefits need to be explicitly compared and uncertainties highlighted.

The case for a combined Environmental Assessment and IPC procedure is a compelling one, especially if 'options' and 'alternatives' and source reduction measures are to be further encouraged. IPC regulators will need to distinguish the phases that exist, over time, in a given industrial plant's waste reduction programme. Different regulations and market-based packages will have to be applied with this phasing in mind.

Finally, while economists have greatly improved the method and techniques necessary to derive monetary values for a range of environmental effects, there are limits beyond which meaningful monetary estimates may not be possible. Multicriteria methods can play a valuable heuristic role in this context.

REFERENCES

Andreottola, G. (1989) A method for the assessment of environmental impact of sanitary landfill, in *Sanitary Landfill: Process, Technology and Environmental Impact* (ed. T.G. Christensen), Academic Press, London.

Barlett, R.V. (1990) Comprehensive environmental decision making: can it work? in *Environmental Policy in the 1990s: Toward a New Agenda* (N.J. Vig and M.E. Kraft), CQ Press, Washington DC, pp. 1–23.

Council on Economic Priorities (CEP) (1986) *Hazardous Waste Management: Reducing the Risk*, Island Press, Washington DC.

Croft, B. and Campbell, D. (1990) *Characterisation of 100 Landfill Sites*, Harwell Waste Management Symposium 1990. Environmental Safety Centre, Harwell, pp. 13–19.

Department of the Environment (DOE) (1986) *Landfilling Wastes*, Waste Management Paper No. 26, HMSO, London.

Fischoff, B., Lichenstein, S., Slovic, P., Derby, S.L. and Keeney, R.L. (1981) *Acceptable Risk*, Cambridge University Press, Cambridge.

James, D. (1985) Environmental economics, industrial process models, and regional-residuals management models, in *Handbook of Natural Resource and Energy Economics* (eds A.V. Kneese and J.L. Sweeney), vol. 1, North Holland, Amsterdam, pp. 271–324.

Kemp, R. *et al.* (1991) *Policy Instruments to Stimulate Cleaner Technology*. Paper presented at the EAERE Conference, Stockholm, June.

Lave, L.B. and Malès, E.H. (1989) At risk: the framework for regulating toxic substances. *Environ. Sci. Technol.*, **23**, 386–91.

Loehr, R.C. (ed.) (1985) *Reducing Hazardous Waste Generation*, National Academy Press, Washington DC.

McAllister, D.M. (1980) *Evaluation in Environmental Planning*, MIT Press, Cambridge, MA.

Maimone, M. (1985) An application of multi-criteria evaluation in assessing municipal solid waste treatment and disposal systems. *Waste Management and Resources*, **3**, 217–31.

Nelson, R. and Winter, S. (1977) In search of a useful theory of innovation. *Res. Pol.*, **6**, 36–76.

Nijkamp, P. (1987) Mobility as a societal value: problems and paradoxes, in *Transportation Planning in a Changing World* (eds P. Nijkamp and S. Reichman), Gower, Aldershot.

Norton, G.A. (1984) *Resource Economics*, Edward Arnold, London.

Organization for European Economic Co-operation (1985) *Environmental Policy and Technical Change*, OECD, Paris.

Organization for European Economic Co-operation (1987) *Environmental Monograph No. 9: The Promotion and Diffusion of Clean Technology in Industry*, OECD, Paris.

Opschoor, J.B. and Pearce, D.W. (eds) (1991) *Persistent Pollutants: Economics and Policy*, Kluwer, Dordrecht.

Opschoor, J.B. and Vos, H.B. (1989) *Application of Economic Instruments for Environmental Protection in OECD Countries*, OECD, Paris.

Parfitt, J. (1990) London's hazardous wastes – a quantitative approach to regulation and planning. PhD Thesis, University of East Anglia, Norwich.

Pearce, D.W. (ed.) (1991) *Blueprint II*, Earthscan, London.

Pearce, D.W. and Nash, C.A. (1981) *The Social Appraisal of Projects*, Macmillan, London.
Pearce, D.W. and Turner, R.K. (1984) The economic evaluation of low and non-waste technologies. *Res. Conserv.*, **11**, 27–43.
Pearce, D.W. and Markandya, A. (1989). *The Benefits of Environmental Policy, Monetary Valuation*, OECD, Paris.
Pearce, D.W., Markandya, A. and Barbier, E.B. (1989) *Blueprint for a Green Economy*, Earthscan, London.
Peãce, D.W. and Turner, R.K. (1990) *Economics of Natural Resources and the Environment*, Harvester Wheatsheaf, Hemel Hempstead.
Pearce, D.W., Barbier, E.B. and Markandya, A. (1990) *Sustainable Development: Economics and Environment in the Third World*, Edward Elgar, Aldershot, Hants.
Pearce, D.W. and Turner, R.K. (1991) The economics of solid waste management: a conceptual overview. CSERGE discussion paper, WM.91–04, University of East Anglia and University College, London.
Powell, J.C. (1991) Approaches to the evaluation of energy from waste systems. PhD Thesis, University of East Anglia, Norwich.
Sobral, M.M. *et al.* (1981) A multi-criteria model for solid waste management. *J. Environ. Management*, **12**, 97–110.
Turner, R.K. (1991a) Municipal solid waste management: an economic perspective, in *The Treatment and Handling of Wastes*, (eds A.D. Bradshaw, T.R.E. Southwood and F. Warner), Chapman and Hall, London, pp. 85–104.
Turner, R.K. (1991b) Benefits of PMP control: ecosystems, in *Persistent Pollutants: Economics and Policy* (eds J.B. Opschoor and D.W. Pearce), Kluwer, Dordrecht.
Turner, R.K. and Bateman, I.J. (1990) *A Critical Review of Monetary Assessment Methods and Techniques*, Environmental Appraisal Group, School of Environmental Sciences, University of East Anglia.
Turner, R.K. and Powell, J.C. (1991) Towards an integrated waste management strategy. *Environ. Management Health*, **2**, 6–12.
Voogd, H. (1983) *Multicriteria Evaluation for Urban and Regional Planning*, Pion, London.
Walshe, G. and Daffern, P. (1990) *Managing Cost Benefit Analysis*, Macmillan, London.
World Commission on Environment and Development (WCED) (1987) *Our Common Future*, Oxford University Press, Oxford.

Case study: industry 13

G. Wyburd

Giles Wyburd has worked in industry, in marketing and general management positions all his life. He was UK Director of the International Chamber of Commerce (ICC) 1983–90. He is now Special Advisor to the ICC.

13.1 INTRODUCTION

Nowhere are environmental dilemmas more acutely felt than in the relationship of industry and business with the rest of society; nowhere are generalizations and over-simplifications about such dilemmas more likely to confuse the issues.

Some critics indeed question how much of industry's activity is necessary or desirable at all. Their idea is to return to some sort of simple preindustrial society, an idea both romantic and elitist (it would only be practicable in the UK if the population were reduced to about ten million) and ignores the aspirations of the majority of city dwellers whose numbers are growing virtually exponentially in both developed and developing countries. To deny on environmental grounds the Chinese people their aspiration to have a fridge in every home would be arrogant and selfish.

The biggest environmental problem of all is that there are so many of us on this earth; the greatest dilemma is that the best way of reducing population growth – short of war, famine and disease – is to raise the standard of living of the poor through economic growth. Chris Patten,

Environmental Dilemmas Ethics and decisions
Edited by R.J. Berry
Published in 1992 by Chapman & Hall, London. ISBN 0 412 39800 1

former British Secretary of State for the Environment, once said that 'poverty is perhaps the greatest pollutant of all' and that 'the idea of zero or negative growth is a council of despair which we must reject' (Patten, 1990).

However, this poses for politicians and industry an enormous challenge, for industry will have to do a much better job in the way it is organized and run than it did in the past. William Blake's 'dark satanic mills' embody concern about the arrogant exploitation of people, raw materials and the environment which the Industrial Revolution brought with it. Today this has left us with a continuing dependence on milder forms of such exploitation and a great deal of mess to clear up. Once it was fashionable to blame the crudest aspects of capitalism for this phenomenon, because the interests of shareholders were paramount; ironically we now see that it is the socialist societies that subordinated the environment and resources of all kinds to the prime aim, namely production. A revealing statement by the British Labour Party in October 1990 was: 'Labour for its part must overcome our traditional image as a producing party, apparently giving priority to jobs and pay packages rather than to environmental concerns' (Labour Party, 1990).

In our search for a lifebelt in this dangerous situation, the concept of sustainable development has been seized upon. Although not invented by the World Commission on Environment and Development (1987), chaired by Gro Harlem Brundtland, the Norwegian Prime Minister, the idea was popularized in the report which bears her name. Brundtland's definition, which was to ensure that development 'meets the needs of the present without compromising the ability of future generations to meet their own needs', was well received by business as it laid to rest the earlier rejection of economic growth by the 'Club of Rome'.

The fact that the concept of sustainability has so many other definitions is both a weakness and a strength. Its weakness is that it can be manipulated to suit every interest, which makes it a favourite with politicians. Frances Cairncross (1991), environmental editor of *The Economist*, is cynical: 'Every environmentally aware politician is in favour of it, a sure sign that they do not understand what it means'. The strength of the concept is that it has brought so many disparate groups to the same table and has produced much-needed common ground between countries in very different stages of economic development, even if there is plenty of room for disagreement on the way as evidenced at UNCED in June 1992.

Certainly the concept of sustainability has underlined the important role of industry in the process of environmental improvement. As Tom Burke of the Green Alliance has put it, 'Our economic and environmental well-being are interdependent . . . responsibility for the environment thus entails concern for industry's success. On the other hand, industry's success may also depend on responsibility for the environment' (Burke and Hill, 1990, p. 4).

13.2 THE 'GREENING' OF ENTERPRISE

An ethical approach to the environment is now accepted by a growing number of responsible companies. In the 1960s enlightened businesses were already developing wider concepts of social responsibility than were written into their articles of association. At that time protection of the environment was not the primary concern. However, the principle that the stakeholders in the company were not just the shareholders, but employees, customers, suppliers and society at large was easily extended to cover environmental considerations. This was indeed so in the case of visionaries such as Tom Watson, the founder of IBM, who believed in quality and excellence in everything the company did. His son, Tom Watson Jr, extended those principles by calling on his managers in 1971 to be continually 'on guard against adversely affecting the environment' (A.B. Cleaver in International Chamber of Commerce, 1990a, p. 54). In the same year the International Chamber of Commerce (ICC), the world's leading business organization spanning all types of business in countries in every stage of development, decided to start formulating ideas, culminating in 1974 in the first *Environmental Guidelines for World Industry*. These were – of necessity – general, providing benchmarks for individual companies to adapt to their own circumstances in their specific industries in each country (ICC, 1990b).

In those early days, there was clearly an element of defensiveness in industry's attitudes. There was fear of excessive or badly focused regulation. There was concern that some such regulation would not be cost-effective and that sensible self-regulation would be preferable. In the early 1980s, industry was emerging from recession and deeply worried about the costs of environmental protection. Those concerns are still real enough. Many companies, the smaller ones in developed countries but the majority of those in the developing world, are more concerned with sheer survival than with wider concepts of responsibility to society.

In the 1980s it fell to the larger companies, especially those in the more obviously vulnerable industries, to give the lead by adopting a proactive attitude. It is they which set the example in developing principles of good environmental management and practice to business as a whole. Neville Cooper, Chairman of the Institute of Business Ethics (IBE), has commented 'It is likely that companies which take the lead will be those which start from a clear sense of what is ethically right, and then go on to use all their available imagination and skills to ensure that their policies lead to commercial and financial success' (Burke and Hill, 1990, p. 1). Here we have the fundamental concept of stewardship, complemented by recognition that overcoming environmental problems is at worst essential for survival, and at best presents business opportunities. For example, the Responsible Care Programme of the UK Chemical Industries Association 'is a concept designed to help the chemical industry to improve its health, safety and environment performance and to enable companies to demonstrate to the public that such improvements are in fact taking place' (Chemical Industries Association, 1990, p. 7).

Sir Trevor Holdsworth, a former president of the Confederation of British Industry (CBI), put it like this: 'The commercial gains enjoyed by companies that pursue environmental excellence are not confined to their current activities. A good company reputation buys credit, not just at the local bank but also with the local community, shareholders and customers. It is created by the quality of goods and services and a company's social standing. Care for the environment is a social responsibility and hence a statement of a company's credit-worthiness and long-term stability' (Elkington, 1990). These examples show how business organizations have, as Tom Burke put it, 'a crucial role to play in fostering corporate responsibility on the environment'. They 'are also a place of peer-group pressure – one of the most effective ways of achieving change' (Burke and Hill, 1990, p. 14).

Responsible business thus recognizes that good environmental practice can be good business and that the influences of all the stakeholders must be heeded. Employees do not need merely to be kept informed but to participate in the development and implementation of environmental policy and practice; indeed some employees are already themselves advocating such policies, while future employees will increasingly take environmental considerations into account before they decide for whom they want to work. Faced with demographic changes which will reduce the number of recruits available, companies ignore this trend at their peril.

The growth of green consumerism, epitomized by sales of over a million copies in more than eight languages of *The Green Consumer Guide* (Elkington and Hailes, 1988) is a clear signal to manufacturers, while 'green chip' security is beginning to emerge as some investors take green considerations into account. Politicians belonging to green parties may not have much chance of assuming power, or be temperamentally suited to doing so, but they are influencing mainstream political parties and governments.

All in all, society at large is exerting stronger pressure on industry and business than ever before to 'do' more for the environment, not recognizing perhaps how much is being 'done' already.

There are two main reasons for this lack of recognition. First, the main 'evidence' that sticks in the mind of the public and others are the disaster stories such as Flixborough, Seveso, Bhopal, the Rhine and Exxon

Valdez. Second, the good news stories, the successes of business, are long term and far less interesting or dramatic than the disasters. Fire prevention seldom hits the headlines; a blazing inferno grabs them.

13.3 THE LESSONS OF ENVIRONMENTAL DISASTERS

A few days after the explosion at the Union Carbide plant in Bhopal, a small group of people from leading UK multinationals was asked to consider the implications. One of their number said: 'We can't discuss it; we don't know what really caused the gas leak'. Another said: 'The cause could be secondary; what matters is that Bhopal will be trotted out over and over again for years to come as an example of the ruthless, uncaring attitude of faceless multinationals operating in the Third World'. This is exactly what has happened. Bhopal, the ghastly tragedy of over 2000 people who died and countless more who were injured, is still the subject of litigation.

The circumstances, including the media reporting of them, were stacked high against objectivity. The Indian Central Bureau of Investigation prevented the Company having access to the plant, records and employees for over a year. The Indian government became a plaintiff in a civil suit against the Company and of necessity fostered a version of the facts that supported its own litigation interest.

When the Company was eventually able to investigate (with all the disadvantages of delay), they concluded that the exceptional entry of a large amount of water into a tank containing methyl isocyanate was the cause of the reaction. They believed this was the act of 'a disgruntled employee' (Kalelkar, 1988); in other words it was an act of sabotage.

Can this be regarded as an act of reckless *corporate* behaviour? As late as the Autumn of 1989, this was the line taken by the campaigning UK consumer magazine *New Consumer* which published a feature article on the 'Corporate (ir)responsibility' of Union Carbide. Whatever is thought about the rights and wrongs of the level of compensation to the victims, there must be a question whether this kind of attack was justified.

However, Union Carbide has not been the only company to suffer because those who find it convenient to argue from the particular to the general have used Bhopal as a stick with which to beat all multinational companies ever since.

The tragedy thus raised many key dilemmas about policy and management. Was it sensible to undertake such a dangerous process as the manufacture of methyl isocyanate in a Third World country? Countries such as India want to manufacture their own chemicals to avoid having to import them. Such decisions are indisputably for them to make.

Was there a design fault that made it possible for such a thing to happen? Was management at fault in terms of supervision? If it was indeed sabotage, was management at fault for driving an employee to do such a thing? Could all this have been avoided if there had been people from the parent company on the spot? India has highly educated managers and understandably limits severely the employment of expatriates.

Why was new housing development allowed by the local authorities so near to the plant after it was built? They must have been aware that the process was potentially dangerous. Yet, with only 50.9% ownership, with no nationals of the USA employed there and no control over the local authorities, the parent company was still largely blamed.

One important dilemma arising from a disaster of this kind results from the complexity of the case and the legal consequences of admitting fault. The subsequent litigation can dilute what must be the prime consideration: concern for the victims. The legal tradition of the USA of going for the 'deepest pockets' regardless of the true facts of the case, and of lawyers being paid by results, is the enemy of honest concern and altruism.

The complexity and bitterness surrounding this and other tragedies and accidents has led to many extreme reactions. Following the Rhine chemical spillages of 1986, Jutta Ditfurth, a prominent German 'green' at that time, said of her country's chemical industry: 'One-third should be kept going, one-third converted to other production – and one-third closed down'. It is unhelpful generalizations of this kind which led Sir Ian MacGregor, a former ICC president as well as one-time chairman of British Coal and British Steel, to express the irritation of industrial leaders faced daily with heavy responsibilities and hard decisions: 'This constant, rather pathetic, longing for simple solutions to complex problems may ultimately undo our society'.

The Green Party has declined in political influence in Germany and elsewhere as its ideas have become absorbed in the mainstream. Tougher government regulation at first raised great concern in German industry that their higher standards would make them uncompetitive with rivals in other countries, such as the UK, where standards were lower. But as management practices improved, they soon realized that in the longer term they would have advantages in meeting wider EC standards, which are often based on their own national legislation.

13.4 INDUSTRY'S CENTRAL ROLE AND ACHIEVEMENTS

German experience has shown that environmentalists certainly have a role to play through their campaigning, but in the end it is industry which can actually *achieve* most through its actions. This led Professor Helmut

Sihler, chief executive of the German chemical company Henkel, to say in 1990 that 'the environment is too important to be left solely to the environmentalists' (International Chamber of Commerce, 1990a, pp. 23–4). By this he meant primarily that the innovation necessary to deal with environmental problems can only come from industry. Peter Bright of Shell International, who has led the ICC's work on sustainable development, added the point about wealth creation: 'Business is often seen as the enemy of the environment, the polluter. There is, however, a tide of change and it is now increasingly being seen as the essential partner, as part of the solution and not the problem, the provider of the wealth and resources needed' (International Chamber of Commerce, 1990a, p. 4).

The President of the ICC in 1989, leading Swedish industrialist Peter Wallenberg, put it like this: 'The onus of proving that sustainable development is feasible rests primarily on the private business sector, as it controls most of the technological and productive capacity needed to conceive more environmentally benign processes, products and services, and to introduce them throughout the world' (International Chamber of Commerce, 1989).

Examples of such achievements, as has been pointed out already, seldom hit the headlines. So far many advances have been achieved with little pain, thereby avoiding the harder dilemmas of the future. One of the best examples is that of the UK chemical industry, which like other energy-intensive industries, has traditionally been concerned with ensuring reliable supplies of the cheapest possible fuels. The problems of acid rain and global warming turned this concern upside down. Robin Paul, chief executive of Albright and Wilson, said in January 1990: 'Industry cannot go on arguing for more and cheaper energy if society as a whole perceives this as encouraging increased energy consumption and causing serious long-term damage to the environment'. He saw the chemical industry's role in this connection as a dual one, namely to use energy efficiently itself and to make energy-efficient products for other industries and for the domestic customer – good housekeeping on the one hand and the positive development of markets on the other.

In the matter of fuel saving, the UK industry's achievement has been impressive (Fig. 13.1). In the 11 years 1967–88, production doubled and energy consumption fell so that the industry used less than half the energy per unit of output than it did before. One of the means by which this was achieved was through innovation – the adoption of combined heat and power schemes generating both electricity and power. Nearly half such schemes in the UK are run within the chemical industry. This is an excellent example of how industry has demonstrated that the quality of life can be improved while energy use is reduced.

Another similar example is to be found in the base metal industries. Conventional copper smelting requires substantial external energy, while

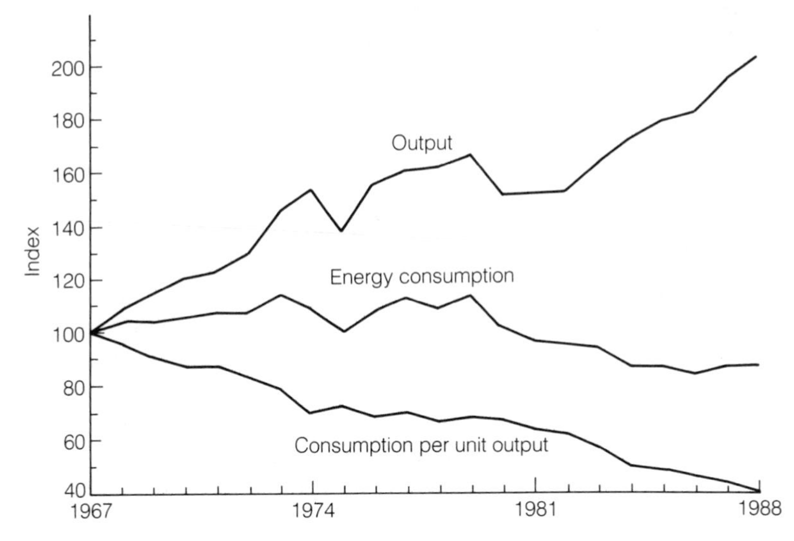

Figure 13.1 UK chemical industry output per energy consumption. From UK Government statistics.

some 70–100% of the sulphur content of the feed is emitted into the atmosphere. Flash smelting was developed by the Outokumpu company in Finland in the 1950s because of energy shortage; it uses the reaction heat produced in the process and no additional fuel is needed. At the same time almost all the sulphur contained in the process feed is recovered into a saleable product: sulphuric acid. The company estimated that some 30% of the world's copper was produced by flash smelting in 1990, showing how much more energy could be saved and pollution reduced as the old processes are phased out (P. Vontilainen in International Chamber of Commerce, 1990a, p. 29). The dilemma here is the non-availability of capital to replace old plant speedily.

Meanwhile the replacement of copper itself in telephone cabling by fibre optics is a further example of the technological breakthroughs that can result in reducing pollution and conserving resources. Such examples are of improvements introduced without particular pressure from environmental considerations.

An example of change introduced purely on environmental grounds is the reduction of phosphates in detergents. Although only about a third of the phosphates causing eutrophication of lakes and rivers comes from detergents, the manufacturers became, understandably, one of the main centres of attention. The German company Henkel, as one of the world leaders in this field, had every incentive to undertake research to find new zeolites to replace phosphates. Their first phosphate-free detergents were introduced in 1986. Before 1980, total consumption of phosphates for detergents in Germany exceeded 250 000 tons; 10 years later the figure was 15 000 tons. Furthermore, Henkel licences have enabled the Japanese to eliminate phosphates from detergents altogether (H. Sihler in International Chamber of Commerce, 1990a, p. 24).

At first sight, this seems simple enough, but it only 'solves' about one third of the problem. What about the other two-thirds, much of which is caused by inadequately treated sewage and agricultural run-offs from fertilizers and animal manure? Remaining producers of phosphates for detergents have pointed out that the reduction of eutrophication has not met earlier expectations. A more effective solution, they say, would be if water authorities and governments would invest in phosphate 'stripping' at sewage treatment works. This would cost money of course, but would be more effective in reducing eutrophication and possibly even allow phosphate-containing detergents to come into their own again, not least because zeolites too have their problems.

Whatever the merits of the case, this example illustrates especially well that few of our problems have simple, straightforward solutions and that governments (and the ultimate payers, the public) have to take great care in the choices they make.

From 1975 to 1987 emissions of sulphur dioxide from Swedish pulp and paper mills decreased by 85% while production increased by 28%. The Swedes ruefully pointed out that the amount of sulphur falling on Sweden from other countries increased considerably in the same period (B. Berggren in International Chamber of Commerce, 1990a, p. 122; see also Chapter 4).

A good example of a combination of environmental need and economic opportunity is to be found in the tyre industry. For many years tyres have been remoulded and retreaded but when this is no longer practicable or safe, disposal is a problem, not least because of the amount of space tyres take up. A pyrolysis plant in the West Midlands can convert each year five million vehicle tyres weighing 50 000 tons into 20 000 tonnes of light fuel oil, 17 000 tonnes of coke and 7000 tonnes of scrap steel (Davis, 1991).

These few examples make far less compelling reading than environmental disasters, but they do show what industry can and will achieve through innovation. Indeed, as Chris Patten said in a speech to the ICC in April 1990: 'Any minister, however ambitious or talented, is far less likely to do as much for the environment as a company making a breakthrough on . . . energy efficiency'.

13.5 REGULATION AND INNOVATION

This illustrates a very real dilemma. While recognizing that regulation is necessary, industry tends to prefer to regulate itself, albeit within a framework of law. This is because external regulation can be heavily bureaucratic, badly focused, and a real burden without necessarily producing the intended results. The more time a company spends on required compliance, the less time and resource it has to innovate. In the end, the argument goes, some innovations can and will leapfrog the more pedestrian effects of regulation which, by definition, is concerned with state of the art industrial activities.

In its briefing for the Heads of Governments meeting at the G7 summit in Paris in 1989, the ICC wrote: 'Business recognizes the need for environmental regulations. But such regulations should provide flexibility and performance orientation for industries and companies to adapt them to their own individual situations. Regulations must not hinder technological innovation which is vital for producing the clean technologies that are the key to sustainable development'.

Industry does, however, deplore regulation that unduly distorts international trade relationships, as the previously mentioned German example has shown. The ideal therefore is for international agreements to be reached not just within a grouping such as the European Community but more widely; the problem Chris Patten has commented, is that the achievement of such agreements make disarmament negotiations appear simple.

Industry has welcomed the increased understanding among environmentalists that extreme demands for regulation can be harmful, just as it has welcomed the greater understanding for the positive contributions that must come from industry and the need for continued, even if more selective, economic growth.

13.6 THE GROWTH OF CONSENSUS

In this respect, responsible industrial leaders and realistic environmentalists have come perceptibly closer to one another in their attitudes. This convergence, which also includes sensible politicians, scientists and trades unionists, was specially notable at the UNECE conference in Bergen in May 1990, where natural differences of views and approach were overlain with a growing understanding of each other's positions and produced an agreed joint agenda for future action.

At Bergen, world industry adopted its own agenda for action and reiterated its commitment to the principle of sustainable development and a decision was made to produce a *Business Charter for Sustainable Development*. This Charter establishes a framework for businesses of every kind, in countries in different stages of development, for environmental improvement; it also demonstrates, to governments and the public, the commitment of international business to that end (Appendix, p. 222).

The Charter was published at the Second World Industry Conference on Environmental Management (WICEM II) in Rotterdam in April 1991 where 750 delegates, including an unusually high proportion of chief executives of leading companies, prepared to put their commitment on the line. By the time of UNCED in June 1992, about 1000 leading companies and business organizations had come out in support of the *Business Charter* and the number continued to grow; leading companies realized that it was not enough to subscribe to principles, they also had to provide examples of those principles in action (International Chamber of Commerce, 1992). Thus it was hardly surprising that the Declaration at the end of WICEM (International Chamber of Commerce, 1991) paid particular attention to the need of business to communicate much more to all its stakeholders about its achievements and future goals. 'Don't trust us, track us' said one, 'watch us as we dedicate ourselves to performance improvement'.

These principles were widely reflected in *Ethics, Environment and the Company*, published by the Institute of Business Ethics (Burke and Hill, 1990) and *Your Business and the Environment*, published by Business in the Environment (1990) under the patronage of the Prince of Wales.

Business in the Environment was set up precisely because such principles of management were still not applied widely enough. It and other bodies are in broad agreement that every company should:

1. Develop a statement of its environmental policy in outline;
2. Undertake a survey of the environmental effect of its whole range of activities, including production processes, R and D, office procedures and transport;
3. Then refine its policy and make it a board-level responsibility to carry it out;
4. Undertake environmental impact assessments (EIA) of all new projects;
5. Undertake regular and rigorous environmental audits of current operations;
6. Assess all new products on the 'cradle-to-grave' basis. This means materials used, design, production process, packaging, use in the hands of the consumer, length of product life, reparability, availability of spare parts and recyclability after use;
7. Train and involve its employees in all aspects of all the above and ensure commitment;
8. Report results, internally and externally, as necessary.

As with all management systems and principles, these constitute no automatic passport to success. If not carried through professionally and with commitment, they merely remain worthy aspirations or exercises in easily-challenged public relations. Even where companies are absolutely serious, they will not necessarily avoid disaster. Would a 1990s-style environmental audit have prevented any of the well-publicized environmental disasters? Not necessarily, but it would have reduced the chances of them happening.

Good environmental management practice may have originated in the most vulnerable industries – chemical, mining, oil and the high-energy industries such as steel, aluminium and glass – but it applies to every type of manufacturing and service industry as well. Indeed, such policies apply to national and local government, health services and education. As an example of the last of these, in October 1990 the Directors of 26 UK polytechnics signed the following declaration: 'This Polytechnic will seek to promote greater awareness of environmental issues through its curricula and endeavour to reduce the damaging environmental impacts of its institutional practice' (Khan, 1990).

Some energy saving, being good housekeeping, can show up directly on the bottom line; some needs investment and requires time before it shows a return. As with most aspects of environmental protection, it is the last 10–20% of the potential saving that costs most to achieve. As the costs related to benefits go up, so the dilemma on what to do increases. Similar principles apply to the reduction of waste. But we must reorganize that, in manufacturing in particular, such reduction is in no way new as a normal function of cost containment.

The goal posts can be moved by external economic forces or by direct government actions. An obvious example of the former is offered by the two oil price rises of the 1970s and the early stages of the Gulf crisis of 1990–91; an example of the latter is the differential taxation in favour of lead-free petrol. As soon as it became demonstrably cheaper, concern about slightly reduced performance and higher consumption was reduced and the market developed rapidly.

As landfill sites for waste disposal become less easily available and less acceptable because of the pollution they cause, so the cost of disposal will rise and it will become more worthwhile to reduce waste yet further and to recycle more.

The work being done by BMW and others on cars designed for their ease of recycling, together with other more conventional requirements, is an example of forward thinking. A car of advanced design which is uneconomical to produce in the early 1990s may prove its worth at the end of the decade. Meanwhile intermediate versions make good advertising . . .

13.7 THE BIG DILEMMAS

So while limitation of waste and good housekeeping generally present few dilemmas to industry, there are many aspects of the environmental debate that seriously challenge it. One of these is the 'precautionary principle'. This is described in the Labour Party document, *An Earthly Chance*, as 'always giving the environment the benefit of the doubt'. The obvious main area to which the principle might be applied is global warming, where there are massive theoretical threats but inadequate information on the timing, nature and extent of climatic change. 'We cannot,' says the document, 'afford to defer hard decisions while calling for further research and waiting for more reports' (Labour Party, 1990). The UK government White Paper (1990) which preceded the Labour document was less open-ended in the matter: 'Action must be based on the best science available, but scientific uncertainty must not be an excuse for delay where there are clear threats of damage that could be serious or irreversible'. Indeed the paper emphasized the importance of resisting populist pressures and dealing with 'fact not fantasy', and 'the need, in environmental decisions as elsewhere, to look at all the facts and likely consequences of actions on the basis of the best scientific evidence available. Precipitate action on the basis of inadequate evidence is the wrong response'.

While understanding the political attractions of the precautionary principle and that the need for further scientific data must not be used as an excuse for doing nothing, industry fears that it could open a veritable pandora's box of ill-focused regulation. It is a classic area where governments, industry and environmentalists need to work together to try and agree in advance on priorities. Having done that, consumers must be made to understand the implications for them in terms of cost.

Speaking of global warming, Patrick Gillam, a managing director of BP, said that the American public was 'not yet ready to accept unlimited extra costs for an environmental trade-off it does not comprehend. This means that if we are to progress we need to solve these dilemmas. Clearly, for example, there is a need for society to be properly informed of the issues and the options, if the larger moves to deal with global warming are ever to become a reality'. He went on to say that 'there are estimates which suggest that simply stabilizing carbon dioxide by the turn of the century will cost European nations 2% of GNP and the USA 5%' (Gillam, 1990).

In other words, cost is not just the dilemma of industry; it applies to all society and above all to consumers buying products and paying taxes. The dilemma for companies is judging how much and when their customers are prepared to pay more. This question raises the biggest dilemma of all, which dominated the Rio Conference in June 1992.

A better environment cannot be obtained on the cheap. While better-off and better-educated consumers may be ready to pay premiums for greener products such as organically grown vegetables, the poor and deprived have other priorities. Ironically, it is the better-off who can best afford to insulate their homes or buy more energy-efficient domestic appliances. Industry itself cannot afford to get too far ahead of market demand, however responsible it may feel. However, it can influence consumers through its marketing; the success of the Body Shop is a classic example.

But such examples are simplistic, in that the consumer decision is clear cut; it is much harder for consumers to realize the enormous costs now facing industry and business and to understand how those costs will inevitably work through into product prices, some over considerable periods of time. Such costs include:

1. High prices of raw materials to replace those no longer acceptable (the most promising CFC substitute costs four times more than CFCs);
2. Investment in new plant to reduce pollution (a horrendous problem for most of industry in what we used to call Eastern Europe);
3. Higher costs of compliance with regulations and monitoring (in the USA some 90% of environmental audits are said to be needed for compliance reasons alone);
4. Levies and charges for authorizations from government (a growing burden under national and EC regulations);
5. Higher environmental insurance premiums (traditionally in the USA such insurance was almost unobtainable, and has proved to be an unbearable burden on some Lloyd's syndicates);
6. Lower sales growth or loss of sales in certain businesses;
7. Financial losses to companies suffering accidents or in breach of regulations. Such losses can include punitive fines, legal charges and consumer boycotts.

The 'polluter pays' principle sounds fine in theory, especially to politicians. The idea, however, that somehow such costs will be borne by somebody else – companies, underwriters or whoever – is a sad example of economic illiteracy. The imposition of strict civil liability in the USA, under its so-called 'superfund' legislation, may not be mirrored in future EC directives, but regulations will still impose tremendous burdens on industry and thence on society. Furthermore, where such legislation is retrospective, as in the USA, making the original polluter bear the costs of cleaning up an industrial site can result in the elimination of that company through the imposition of today's standards. Even worse in a sense, it can destroy a company that has unwisely bought the site in question without proper survey.

It is therefore no surprise at all that the larger companies in vulnerable industries have grown in awareness of the need to develop environmental policies and preventative management systems. It should also be a subject of increasing concern to those which have done little so far in this respect. At the very least, they should realize that, quite apart from moral considerations, prevention usually costs less than cleaning up at a later date.

This is where the leadership shown by the more aware companies assumes tremendous importance. A senior UK manager once queried why he should take time to speak at a conference on environmental audit through which others would benefit from his company's expertise. He was soon persuaded to see that his company not only had a duty to take on the task, but would also derive benefit by helping to raise industry's general awareness of the tasks it faces. IBM, under the leadership of the Watsons already mentioned, is a very good example. The Chairman and Chief Executive of IBM UK, Tony Cleaver, in his speech at Bergen in May 1990, explained that the company not only had its internal environmental policies, but was specifically committed to articulating its vision to the outside world. Thus it had undertaken to eliminate CFCs from its products and processes by the end of 1993, well ahead of international agreements enshrined in the Montreal Protocol. This was involving expenditure both in 'considerable' investment and high material costs. The commitment to the outside world involved giving employees and pensioners financial support to help them in external environmental work. His own personal contribution was to head the Business in the Environment target team which made its first presentation to the public in November 1990. Policies such as those shown by IBM are clearly a very balanced combination of fundamental responsibility and business sense.

The advantages or disadvantages of getting too far ahead of regulations clearly need to be carefully weighed. At the same time, industry's concern for a 'level playing field', i.e. national or regional regulations not too far ahead of competitors, can prove to work both ways, as the case of the Germans' regulation of their chemical industry has already shown. The German waste disposal industry was much larger than that of the UK in 1990, but included a 30% export element, as against only 5% in the smaller UK industry. It is of little comfort to business and consumers in Europe that taxes on petrol are so much higher than in the USA, where transport costs are therefore lower, while the USA's contribution to greenhouse gas emissions is so high. Europeans can be forgiven for finding the Californian vote to increase gasoline prices by ludicrously small amounts over extended periods as tinkering at the edges of the problem. But California is already leading the world in reducing emissions rather than motoring. Some 2% of cars sold in 1998, rising to 10% in 2003, must be 'zero emission vehicles', presumably electric. It will be fascinating to see if such regulations give the USA a lead in cost-effective electric vehicle technology, and whether these modest targets will be raised. Yet there is a dilemma here too; electricity is at present a greater contributor to carbon dioxide emission than oil, the argument goes, so vehicles with rechargeable batteries may not be as benign as claimed.

There is a parallel here with the dilemma about nuclear power (Chapter 6). Certainly the arguments and counter-arguments generate more heat than light. In favour of nuclear power is the fact that present high relative costs will reduce as fossil fuels become less available and more expensive. This, added to the elimination of carbon dioxide emissions, makes the argument powerful. The counter-arguments about safety (epitomized by Chernobyl), about waste disposal and about nuclear weapons as a byproduct are voiced with understandable passion by opponents of nuclear power.

The need is for further innovation to overcome these problems to the extent that nuclear power ceases to pose the dilemma it does now. Nowhere is there a greater need for less posturing on both sides of the argument and more consensus on what solutions might be more generally acceptable.

The nuclear debate pre-eminently illustrates the dilemma of sustainable development in relation to the threat to future generations. It is also a stark reminder that the idea of avoiding every conceivable risk is far from simple.

Such questions are of special importance to developing countries, where we are faced in the 1990s with the greatest North–South issue of all, namely how to help them raise their standard of living and reduce population growth without destroying the planet. The Chinese aspiration to have a fridge in every home has already been mentioned. Extend that idea to enabling only the Chinese and the Indians to have, say, *half* the number of cars per head than the people of the EC would already mean another 380 million cars in circulation (Table 13.1). The implications for the environment are clear enough.

The Chinese are eloquent on this point. Their representative at the 1989 London conference on the ozone layer made it clear that developed countries had enriched themselves at the expense of the planet, and that therefore they owed it to the developing world to give substantial help to raise the standard of living on the basis of new and cleaner technology.

This poses the political dilemma which hung like a cloud over Rio, as it

Table 13.1 Approximate car density in 1990

	Population (millions)	Cars in Use (millions)	Cars per 1000 People
European Community	322	126	390
India	810	1.65	2
China	1160	1.1	1

Source: SMM&T

as it implies a need for aid budgets well above the present UN target of 0.7% of gross national product, which few countries currently meet. For industry it poses tremendous dilemmas relating to the protection of intellectual property rights. Patentable processes, however strong the moral imperatives, must not be given away if research and innovation are to continue. A further question is the suitability of potentially dangerous processes to developing countries. CFC replacements, which they need very urgently, present an obvious example. The lessons of Bhopal may again become all too relevant.

Much can be done for the Third World by the adoption of the appropriate intermediate technology so strongly advocated by E.F. Schumacher. There are many examples of this; an excellent one is roofing in West Africa. From the 1930s onwards it was a status symbol to have corrugated iron roofs on houses and small buildings. This was good business for the steel industry in South Wales before massive Japanese competition took it away. Now a small British company called J.P.M. Parry and Associates has won an award from the UK Centre for World Development Education (1990) for developing a process and equipment to make lightweight tiles of a mixture of cement, sand and chopped fibre. The tiles are much lighter than clay tiles, thus needing less stout timber support, and can be made from locally available materials. They are more environmentally friendly in production and use than corrugated iron, which costs twice as much, and are being made in over 50 countries.

13.8 CONCLUSION

While the major economic, political and technological dilemmas relating to economic growth and the cost of improved environmental policies in countries at all stages of development are as great as any that the world has ever faced, there are grounds for hope.

The increased recognition that everyone, from the chairman of leading multinationals to Brazilian peasants, is in this together, is just beginning to concentrate minds. Businessmen, environmentalists and politicians are finding increased common ground.

The capabilities of industry at all levels of technology to find the answers are increasingly recognized. Industry, for its part, is becoming less defensive and recognizing its responsibilities, while employees at all levels are becoming involved. They need to be, because they will be faced with more and more environmental dilemmas, and more and more hard choices on which the very future existence of their companies may depend.

They are also having to be much less defensive towards the public at large, who are also their customers and who demand to know more and more about what they do and what they achieve in improving environmental performance. Frank Popoff, Chief Executive Officer of the American company Dow Chemical, said at the ICC Congress in Hamburg in 1990: 'They (the public) do not want to know how much we *know*. They want to know how much we *care*'. He capped that by urging his hearers to avoid 'greenwash', and not to be 'more vocal than we are committed'.

REFERENCES

Burke, T. and Hill, J. (1990) *Ethics, Environment and the Company*, Institute of Business Ethics, London.

Business in the Environment (1990) *Your Business and the Environment*, ICC, Paris.

Cairncross, F. (1991) *Costing the Earth*, The Economist Books, London.

Chemical Industries Association (1990) *Chemicals in a 'Green' World*, London.

Davis, J. (1991) *Greening Business*, Blackwell, Oxford.

Elkington, J. (1990) *The Environmental Audit. A Green Filter for Company Policies, Plants, Processes and Products*. SustainAbility, London.

Elkington, J. and Hailes, J. (1988) *The Green Consumer Guide*, Gollancz, London.

Gillam, P. (1990) *Symposium on Global Warming – no Easy Panaceas*, at Sussex University on 14 November.

International Chamber of Commerce (1989) *Sustainable Development – The Business Approach*, Paris.

International Chamber of Commerce (1990a) *The Greening of Enterprise*, Paris.

International Chamber of Commerce (1990b) *Environmental Guidelines for World Industry*, Paris.

International Chamber of Commerce (1991) *WICEM II, Conference Report and Background Papers*, Paris.

International Chamber of Commerce (1992) *From Ideas to Action*, Paris.

Kalelkar, A.S. (1988) *Investigation of Large Magnitude Incidents: Bhopal as a Case Study*. Arthur D. Little, Cambridge, Massachusetts.

Khan, S.A. (1990) *Greening Polytechnics*, Committee of Directors of Polytechnics, London.

Labour Party (1990) *An Earthly Chance*, London.

Patten, C. (1990) Wilson Lecture, Godolphin and Latimer School, Hammersmith, published in *The Times* 13 March 1990.

White Paper (1990) *This Common Inheritance*, HMSO, London.

World Commission on Environment and Development (1987) *Our Common Future*, Oxford University Press, Oxford.

APPENDIX ICC 'Business Charter for sustainable development: Principles for Environmental Management'

INTRODUCTION

Sustainable development involves meeting the needs of the present without compromising the ability of future generations to meet their own needs. Economic growth provides the conditions in which protection of the environment can best be achieved and environmental protection, in balance with other human goals, is necessary to achieve growth that is sustainable.

In turn, versatile, dynamic, responsive and profitable businesses are required as the driving force for sustainable economic development and for providing managerial, technical and financial resources to contribute to the resolution of environmental challenges. Market economies, characterized by entrepreneurial initiatives, are essential to achieving this.

Business thus shares the view that there should be a common goal, not a conflict, between economic development and environmental protection, both now and for future generations. Making market forces work in this way to protect and improve the quality of the environment – with the help of performance-based standards and judicious use of economic instruments in a harmonious regulatory framework – is one of the greatest challenges that the world faces in the next decade.

The 1987 report of the World Commission on Environment and Development, *Our Common Future*, expresses the same challenge and calls on the cooperation of business in tackling it. To this end, business leaders have launched actions in their individual enterprises as well as through sectorial and cross-sectorial associations.

In order that more businesses join this effort and that their environmental performance continues to improve, the International Chamber of Commerce hereby calls upon enterprises and their associations to use the following Principles as a basis for pursuing such improvement and to express publicly their support for them. Individual programmes developed to implement these principles will reflect the wide diversity among enterprises in size and function.

The objective is that the widest range of enterprises commit themselves to improving their environmental performance in accordance with these principles, to having in place management practices to effect such improvement, to measuring their progress, and to reporting this progress as appropriate internally and externally.

Note The term environment as used in this document also refers to environmentally related aspects of health, safety and product stewardship.

PRINCIPLES

1. *Corporate priority* To recognize environmental management as among the highest corporate priorities and as a key determinant to sustainable development; to establish policies, programmes and practices for conducting operations in an environmentally sound manner;
2. *Integrated management* To integrate these policies, programmes and practices fully into each business as an essential element of management in all its functions;
3. *Process of improvement* To continue to improve corporate policies, programmes and environmental performance, taking into account technical developments, scientific understanding, consumer needs and community expectations, with legal regulations as a starting point; and to apply the same environmental criteria internationally;
4. *Employee education* To educate, train and motivate employees to conduct their activities in an environmentally responsible manner;
5. *Prior assessment* To assess environmental impacts before starting a new activity or project and before decommissioning a facility or leaving a site;
6. *Products and services* To develop and provide products or services that have no undue environmental impact and are safe in their intended use, that are efficient in their consumption of energy and natural resources, and that can be recycled, reused, or disposed of safely;
7. *Customer advice* To advise, and where relevant educate, customers, distributors and the public in the safe use, transportation, storage and disposal of products provided; and to apply similar considerations to the provision of services;
8. *Facilities and operations* To develop, design and operate facilities and conduct activities taking into consideration the efficient use of energy and materials, the sustainable use of renewable resources, the minimization of adverse environmental impact and waste generation, and the safe and responsible disposal of residual wastes;
9. *Research* To conduct or support research on the environmental impacts of raw materials, products, processes, emissions and wastes associated with the enterprise and on the means of minimizing such adverse impacts;
10. *Precautionary approach* To modify the manufacture, marketing or use of products or services or the conduct of activities, consistent with scientific and technical understanding, to prevent serious or irreversible environmental degradation;
11. *Contractors and suppliers* To promote the adoption of these principles by contractors acting on behalf of the enterprise, encouraging and,

where appropriate, requiring improvements in their practices to make them consistent with those of the enterprise; and to encourage the wider adoption of these principles by suppliers.

12. *Emergency preparedness* To develop and maintain, where significant hazards exist, emergency preparedness plans in conjunction with the emergency services, relevant authorities and the local community, recognizing potential transboundary impacts;
13. *Transfer of technology* To contribute to the transfer of environmentally sound technology and management methods throughout the industrial and public sectors;
14. *Contributing to the common effort* To contribute to the development of public policy and to business, governmental and intergovernmental programmes and educational initiatives that will enhance environmental awareness and protection;
15. *Openness to concerns* To foster openness and dialogue with employees and the public, anticipating and responding to their concerns about the potential hazards and impacts of operations, products, wastes or services, including those of transboundary or global significance;
16. *Compliance and reporting* To measure environmental performance; to conduct regular environmental audits and assessments of complianc with company requirements, legal requirements and these principles and periodically to provide appropriate information to the Board of Directors, shareholders, employees, the authorities and the public.

The Business Charter for Sustainable Development was adopted by the 64th Session of the ICC Executive Board on 27 November 1990, and first published in April 1991.

Case study: the Government sector

14

D.A. Everest

David Everest is a chemist, who has spent most of his working life in Government institutions (National Chemical Laboratory, 1956–64; National Physical Laboratory, 1964–77). He joined the Department of the Environment in 1979, retiring as Chief Scientific Officer Environmental Pollution, and Director of the Civil Service Year Book.

14.1 INTRODUCTION

Public concern regarding the protection and improvement of the environment* is a growing political pressure concentrated mainly, but not exclusively, in the economically advanced countries, with environmental protection increasingly being presented as a moral issue. This viewpoint is reflected in the UK Government environment White Paper (1990) which states (para. 1.14) that 'the starting point in this Government is the ethical imperative of stewardship which must underline all environmental policies', and that 'we have a moral duty to look after our planet and to hand it on in good order to future generations'. These principles are expressed in the term 'sustainable development': not sacrificing tomorrow's development for an often illusory gain today. The Government considers that to achieve this end required the full integration of environmental considerations into economic policy, and that economic development and environmental protection need not be irrevocably

* The term 'environment' is taken as comprising land, air and water and the living organisms, including other humans, supported by those media (see Environmental Protection Act, Section 84 (2)).

Environmental Dilemmas Ethics and decisions
Edited by R.J. Berry
Published in 1992 by Chapman & Hall, London. ISBN 0 412 39800 1

opposing principles. Development provides the wealth to invest in cleaner methods of production and to husband natural resources.

A clean and pleasant environment, like a satisfactory level of nutrition, housing, education or employment, is a good generally desired by society. Thus it would be expected that governments will endeavour to promote the right conditions for achieving and maintaining a good environment, although a balance must be struck with the resource demands of other goods desired by the population. How this balance is struck will reflect the political assessment of the relative importance of these different needs and, in a democracy, will reflect the views, values and perceptions of society both generally and with respect to environmental issues*.

However, only rarely are environmental issues of such certainty, clarity, and simplicity that definite decisions and actions can be rapidly taken. It is often difficult for governments to decide where the balance of public perception actually lies because it is usually the actions of a committed minority who win media attention and initially promote particular issues. Sometimes such issues grow into matters of wider public concern demanding political action; on other occasions fade as they are perceived to be of less importance. Thus it is necessary for a government to estimate the importance in environmental and political terms of particular issues, and then to monitor how scientific understanding and public perception develop, defining any need for additional information. When issues start to emerge as being of sufficient importance to warrant action, then the government must consider not only possible control measures but also their impact on the economy and on sections of society whose interests may be affected.

14.2 PRESSURES ACTING ON UK POLICY-MAKERS

Pressures falling on Government policy-makers for action on environmental issues from various sources, reflect a perceived balance between the likelihood and the uncertainties arising from the cause and effect relationships of particular environmental problems. Pressure for action will come from the environmentally-concerned public, both domestically

* It is a basic assumption of this paper that in a democracy, such as the UK, the political parties operating the processes of government attempt to make the best decisions they can for society generally, in the light of their own views and convictions of the manner in which they wish society to evolve. As the winning and retention of power is an essential condition for promoting their aims, it is necessary for the policies of both government and opposition to reflect, and be influenced by, public perception of the issues on the political agenda. This is particularly true for environmental issues where public concern and demands for action are growing and spreading across party lines. This is leading to a growing political consensus for action, although the details of the environmental protection measures proposed are likely to reflect the basic political philosophy of the party in power.

and internationally, and from the understanding of the issues as articulated by the scientific community.

14.2.1 PRESSURES FROM THE ENVIRONMENTALLY-CONCERNED PUBLIC

Pressure for action from the environmentally-concerned public is a potent driving force for Government action on environmental issues. It is often channelled through, and focused by established environmental groups such as the Royal Society for the Protection of Birds (RSPB), the National Society for Clean Air, the National Trust and the National Trust for Scotland, or the Councils for the Preservation of Rural England, Wales and Scotland. These are well established and prestigious organizations, and form an important part of the environmental 'establishment' whose views receive serious attention from the Government. Sometimes the drive for action comes from general environmental pressure groups, such as Friends of the Earth or Greenpeace, with traditions for active propaganda and direct action which can lead to confrontation. Recently the UK Government has endeavoured to build bridges to such organizations by encouraging them to give formal evidence to Parliamentary Select Committees or the Royal Commission on Environmental Pollution, or by directly asking for their views on particular issues (e.g. through the National Council for Voluntary Organizations: White Paper (1990) para. 18.34). In this way these bodies may influence government environmental policy.

Single issue environmental pressure groups often emerge as vehicles for influencing the Government on particular issues. Examples are CLEAR (the Campaign for Lead-free Air), and local or regional groups which oppose specific developments (e.g. new roads, railways, radioactive waste disposal sites, etc.). Often very strong feelings are aroused over such issues with protagonists taking quite radical measures to press their case, including action which others might consider unethical (e.g. leaking confidential documents to the press).

The media plays a significant role in focusing public concern on environmental questions, both for bringing public opinion to bear on Government and also as a means by which the government can influence public perceptions and actions on particular issues. The media often show considerable presentational skills in setting out key aspects of environmental questions but tend to be less effective in presenting the reservations in understanding that scientists often express on environmental problems; this can be a serious difficulty for the policy-maker in both for the formulation and the communication of environmental policy.

Public opinion, aided and focused by the media, is having an increasing commercial impact on the attitude of industry to environment issues.

Examples are the attention being given by the advertising media to the promotion of 'ozone friendly products' (i.e. products which do not utilize chlorofluorocarbons (CFCs) in their use or manufacture*), and the pressures on oil companies to adopt effective safeguards against pollution, and to pay for any clearing up if their safeguards prove inadequate. Indeed public and consumer opinion is becoming an effective market instrument for the promotion of environmental issues, and one which the Government is using to help generate public acceptance for its environmental policies and decisions.

14.2.2 INTERNATIONAL PRESSURES

Increasingly, UK environmental policy is being influenced by forces outside the UK, in particular by our European partners acting through the EC, and by the international community acting through the UN Commission for Europe or the United Nations Environmental Programme (UNEP). This trend is most immediately apparent for global environmental issues, such as protection of the stratospheric ozone layer or climate warming from the greenhouse effect, where action by the UK alone would be ineffective. Development of policy on these problems takes the form of diplomatically-negotiated conventions and protocols under the auspices of UNEP (e.g. the 1987 Montreal Protocol on the control of emissions of CFCs). These negotiations include the scientific, economic and political aspects of the environmental issues being considered. If agreement is to be achieved, they must cover also the differing perceptions and priorities of the economically advanced and lesser developed countries on environmental issues. In particular the lesser developed countries will insist that any measures proposed for global environmental protection must not impede their economic development, as occurred at the 1992 United Nations Conference on Environment and Development (UNCED).

The increasing importance of legislation in the environmental field, both for eliminating barriers to trade as well as for promoting environmental benefit, is an even greater restriction on the freedom of the UK Government to make its own decisions on environmental policy. This transfer of authority from London to Brussels on environmental issues has been reinforced since the implementation of the Single European Act, which involves both majority voting in the Council of Ministers and a greater role for the European Parliament on environmental policy and

* Chlorofluorocarbons (CFCs) act to reduce the amount of ozone in the stratosphere. Because ozone absorbs in the ultraviolet, any decrease in stratospheric ozone will increase the amount of damaging ultraviolet radiation reaching the earth's surface. It is likely that their use will be phased out by the year 2000 (White Paper, 1990). The CFCs are also greenhouse gases, so that eliminating their emissions will act to limit a greenhouse effect-related climate warming.

legislation. As a result, UK environmental policy will be increasingly influenced by the perceptions and concerns of the wider European public on environmental questions, which often differ from those of the UK public. As discussed later, these differences in perception over the perceived effects of acid precipitation and photochemical air pollution on European rivers, lakes and forests have led to the UK being labelled an environmental 'laggard', or even as the 'dirty man of Europe' by its European neighbours (Brackley, 1987; Chapter 4).

14.2.3 THE INFLUENCE OF THE SCIENTIFIC COMMUNITY

In the environmental field the scientific community can be a potent source of pressure on government for action. Their views are particularly influential in those countries, such as the UK, where the stated aim of government is to develop rational environmental policies based on considered scientific analysis of the relevant cause and effect relationships. The problem here is that the relevant cause and effect relationships are not only complex in character, bordering on the margins of scientific knowledge, but may indeed transgress the boundaries of the scientific method itself. In these circumstances the views of the scientific community have a substantial influence on policy, not only by directly influencing the views of Ministers but also the views of those who have the responsibility for advising Ministers on the balance of risk of potential adverse environmental impacts and the political and economic costs of possible counter measures.

14.3 THE NATURE OF EXPERT SCIENTIFIC ADVICE

Much of the expert advice given to environmental policy-makers in government covers formal scientific analysis. However, it must be recognized that decisions on environmental questions involve political, social and economic considerations in addition to the underlying scientific aspects of the problem. Indeed, many environmental issues, notably those associated with the term 'risk', must be classed as being 'trans-scientific' in character. The latter term means that that although in principle these problems are amenable to the scientific approach they cannot be fully answered by science*, and must incorporate an essential element of public and political judgement (Everest, 1990).

* As discussed by Weinberg (1972), the term 'trans-scientific' can be used in three rather different senses. In the first (low level insult), science is inadequate because to get answers would be impractically expensive. Secondly, in the area of the social sciences the subject matter is usually too variable to allow rationalization according to the strict scientific canons established in the natural sciences. In the third case (choice in science), science is inadequate because the issues themselves involve moral and aesthetic judgements: they deal not with what is true but with what is valuable.

In principle any system of providing advice to government on environmental matters will have to consider also the trans-scientific aspects of particular environmental issues. Such trans-scientific problems require a combined approach by both natural and social scientists. The former use conventional methods of logical positivistic analysis, while the latter use behavioural and cognitive approaches as to how people make judgements. Generally environmental risk issues involve a fusion of the scientific and social science approaches on the grounds that a technical hazard cannot be fully evaluated unless it is judged in its social context (Douglas, 1987).

A frequently encountered question concerns the extent to which precautionary action is justified before firm scientific understanding has developed. The case for precautionary action is that once environmental damage has arisen it is often irreversible or slow to rectify so justifying immediate counter action; it is not safe to await the relatively slow evolution of a balanced scientific understanding of the issue. On the other hand, developing scientific understanding of the problem may show that these preventative measures were ill-directed and that the resources utilized would have been more usefully applied elsewhere. This is an area where the policy-maker needs expert guidance and advice when deciding the case for preventative action, noting also that future technological developments may well cheapen any environmental protection measures, so reducing the economic risks of preventative action.

14.3.1 THE PROVISION OF EXPERT ADVICE TO GOVERNMENT ON ENVIRONMENTAL ISSUES

Reliable and authoritative expert advice is an essential ingredient in the development of policy. In the UK expert environmental advice available to the government includes the internal resources in Government Departments and their associated Research Establishments, supported and reinforced by the environmental research funded by these Departments, and by publicly funded bodies such as the Natural Environment Research Council (NERC) (Annual Review of Government funded R and D, 1989). Study groups especially established to cover specific areas are another important source of advice, e.g. the Review Group on the stratospheric ozone problem (DOE, 1988).

A significant source of advice comes from formal standing advisory committees which play a particularly important role in the environment field. Advice provided by expert civil servants or by a wider scientific community is of great utility, but cannot replace the authoritative policy-orientated advice which a well-established expert committee can provide. These committees can express a consensus view of its members on contentious isues, so avoiding the reservations expressed by individual scientists.

Examples of advisory committees in the environmental sector are the Royal Commission on Environmental Pollution (RCEP), the Advisory Committee on Pesticides (ACP), the Radioactive Waste Management Advisory Committee (RWMAC), and the Committees on the Medical Aspects of the Contamination of Air, Soil and Water (CASW) and on the Medical Aspects of Radiation (COMARE). These bodies cover subject areas of environmental importance and of actual or potential political sensitivity, although environmental considerations are not necessarily their primary concern.

Although the RCEP is predominantly a scientific body, it has members with expertise in law, economics and business, and is thus well placed to bridge the gap between the views of the scientific community and practical policy-making. The RCEP is independent of government, in particular as regards the selection of study topics but its reports are given a formal government response. These reports are the most important output of the RCEP, examples being those on environmental lead (RCEP, 1983) and the release of genetically-modified organisms to the environment (RCEP, 1989).

The ACP is an example of an advisory committee with a particular legislative or regulatory function, namely to advise Ministers regarding approvals of pesticides required under Part III of the 1985 Food and Environmental Protection Act. It is supported by a technical secretariat which provides it with summaries of the extensive technical dossiers supplied by applicants for pesticide approval; information in these summaries is publicly available.

RWMAC, CASW and COMARE are examples of Departmental Advisory Committees which provide expert advice aimed at issues of general or specific relevance to Departments. The task of the RWMAC is to advise Ministers on the management and treatment of wastes from all stages in the nuclear fuel cycle, including the long-term management of spent fuel and the wastes arising from the Fast Reactor; it also endeavours to inform and take account of public preceptions in this field. The views and advice of RWMAC are made public mainly through its Annual Reports, published by HMSO starting in 1980.

In Whitehall it is the responsibility of the Government's Chief Medical Officer (CMO) to advise Ministers (strictly English Ministers) on present or potential health problems, including those concerned with environmental health. The CMO is assisted in this task by a family of specialist medical advisory committees, of which CASW and COMARE are examples, together with other committees covering the medical aspects of food, carcinogenicity and mutagenicity. CASW is particularly concerned with the environmental health effects of chemical pollutants in air, soil and water. As advice on these issues is formally provided by the CMO, the activities of CASW are not usually visible to the general public.

COMARE's inquiries have hitherto been confined to a single issue of considerable public sensitivity, namely the link between childhood leukaemia clusters and nuclear installations. A special feature of COMARE's activities is that it covers an area of scientific uncertainty, a result of incompatible conclusions from epidemiological and radiological approaches (COMARE, 1986, 1988). These special features have resulted in COMARE having greater public visiblility than other CMO committees.

14.3.2 COMMITTEE EVOLUTION

Environmental risk advisory committees tend to evolve through at least two stages, with a third potential evolutionary stage, reflecting that increasingly they are required to consider the trans-scientific aspects of environmental problems (Everest, 1990). In the first stage they restrict themselves to pure scientific advice, aiming to reflect the state of knowledge and to ensure that this position is faithfully transmitted to decision makers; such committees are composed entirely of reputable scientists with a role reflecting primarily legitimacy and authority. In the second stage the committee operates in the trans-scientific mode, and contains members prepared (or expected) to voice a 'public view' of the perspectives and likely reactions of particular groups and to advise how information should be disseminated to help the public become better informed and educated about the relevant state of science. In the third evolutionary stage the committee would adopt a more active public awareness role through balanced provision of thoughtfully judged information; it seems unlikely that this stage will be reached fully by a UK environmental advisory committee in the system as currently structured.

For advisory committees operating in the trans-scientific mode there should be a sequence of authority ranging from scientific probity, through the mediation of contentious scientific dispute to an engagement with the general public in the policy-related issues that emerge. The RCEP seems to be nearest to meeting these criteria and can be considered as a model stage two body. It also has some stage three characteristics, as its reports are both treatises and recommendatory.

The ACP and CASW operate as essentially as stage one bodies although the 1985 Food and Environmental Protection Act will increasingly require the former to pay attention to external views in the advice it gives to Ministers. COMARE has a professional membership and endeavours to confine itself only to giving stage one professional advice, but it is being forced to make a more public interest view because the area it covers is both of public sensitivity and scientific uncertainty with trans-scientific overtones. RWMAC has some members representing the wider public interest; it has a tradition of giving outspoken policy comment in its Annual Reports and it is endeavouring to develop a wider dialogue with the environmental movement. It is thus evolving from a stage one to stage two body, with some indications of moving someway towards stage three status.

14.3.3 INTERNATIONAL ADVISORY COMMITTEES

Similar scientific advisory committee systems exist in some other countries, although a supranational advisory committee system has yet to evolve. In the EC, policy-making in the environmental field mainly reflects political interaction between the Commission, National Governments and the European Parliament. Advice on scientific or technological aspects of these issues is provided by meetings of technical experts, consisting either of officials from the member states, or by consultants funded by the Commission. If the Community evolves into a more politically united body then the setting up of a Europe-wide scientific advisory committee system would appear desirable.

Considerable progress has been made in establishing expert advisory committees for global environmental issues. An example is the Intergovernmental Panel on Climate Change (IPCC), operating under the auspices of the United Nations Environment programme (UNEP) and the World Meterological Organization. The IPCC Working Group 1, which has recently published its scientific assessment of future greenhouse effect-induced climate change (IPCC, 1990), is made up of active working scientists drawn from a wide range of nations and geographical regions. Effectively IPCC-WG 1 is acting as an expert scientific advisory group for the international community. Reports are due to be published by Working Group 2 on the potential impacts of climate change and from Working Group 3 on policy. Other examples of international expert advisory committees are the *ad hoc* bodies which advise UNEP and the cooperating nations on scientific questions relating to the convention and protocol for the protection of the stratospheric ozone layer.

14.4 EXAMPLES OF ENVIRONMENTAL DECISION-MAKING

The following examples are given as illustrations of environmental decision-making. The first is an issue on which action has already been taken, while the others represent issues at differing stages of policy development.

14.4.1 LEAD IN PETROL

Lead and its compounds are known toxins widely used in industry and in domestic applications. Concern has relatively recently arisen over the environmental health effects of long-term exposure to low levels of lead (below those usually considered toxic), particularly in relation to possible neuropsychological effects in children (Rutter, 1983; Medical Research Council, 1985, 1988). There are a number of pathways by which lead reaches human beings (Davies *et al.*, 1990), such as from lead plumbed water supplies, old lead paint and the dust to which it gives rise, industrial emissions, and the use of lead additives in petrol. The last was the subject of intense political debate during the 1970s to early 1980s, with the environmental group CLEAR mounting a strong campaign to remove lead from petrol.

In an endeavour to keep lead emissions constant while petrol consumption was rising, the UK Government made a number of small cuts in the permitted level of lead in petrol from 1973 onwards, culminating in 1981 in the decision to make a substantial reduction from 0.4 to 0.15 g/l, to take effect from the start of 1986. However, these measures were not sufficient to allay public concern and, following a recommendation from the Royal Commission on Environmental Pollution (RCEP, 1983), the UK Government agreed to pursue a policy of removing lead from petrol. In mainland Europe there was less public concern over the environmental health impacts of lead but greater concern regarding the links between photochemical pollution and the increasing occurence of forest damage. As the use of car exhaust catalysts was the favoured method in Europe for reducing their polluting emissions, and as these catalysts require the use of lead-free petrol, there was European support for the removal of lead from petrol, and this is now agreed EC policy.

The effect of lowering lead concentration in petrol from 0.4 to 0.15 g/l was a 50% reduction in air lead concentrations, 16% in blood lead levels for children and 9–10% for adults (Quinn and Delves, 1989). These reductions have to be considered in the content of a long-term fall in blood lead concentrations of about 4–5% a year, presumably from other pathways by which lead formerly reached adults or children. This downward trend is probably more significant for reducing lead body burden than the once for all benefit achieved by removing lead from petrol.

14.4.2 ACIDIC EMISSIONS AND PHOTOCHEMICAL POLLUTANTS

These environmental problems are associated with the primary pollutants in fossil fuel combustion, notably sulphur dioxide, NO_x ($NO + NO_2$) and carbon monoxide, and the secondary pollutants formed from them by atmospheric chemical processes. If emissions come from high stacks, as commonly occurs from fixed emission sources such as power stations, then local ground level pollutant concentrations are minimized. However, emission plumes can travel long distances, allowing time for sulphur dioxide and NO_x to be oxidized to sulphuric and nitric acids. These acids are eventually washed out in rain ('acid rain'), often in regions remote from the original emissions. Photochemical pollutants, of which ozone is the most important, are more often associated with vehicle emissions; these contain the NO_x and volatile hydrocarbon (VOCs) which are essential links in their atmospheric photochemical formation. As vehicle emissions occur at low level, photochemical pollution is more local in character than acid rain, although such pollutants can be transported over large distances under some atmospheric conditions.

Acid rain, to a lesser extent photochemical pollution, are examples of pollution occuring in regions remote from where primary emissions arise.* In such situations, the development of viable control policies must reconcile the inevitable differences in perception between those who suffer damage and those who are obliged to take preventative action. As noted earlier, the former usually emphasize the need for early action, often on a precautionary basis; the latter often resist the adoption of what they perceive to be premature and expensive control measures, and emphasize the need for more research before taking action and thus closing options.

In any negotiations, an 'emitting' country, such as the UK, must consider the consequences of proposed environmental regulations on industry (which in the source of the wealth necessary to generate environmental improvement), on employment and, sometimes, on other environmental issues. Thus an EC Directive (EEC, 1988) on the reduction of sulphur dioxide emissions from large combustion plants (such as UK coal-burning power stations), would require the installation of flue gas desulphurization (FGD) equipment, probably utilizing a lime scrubbing process. This would increase the cost of electricity but also necessitate the use of large quantities of limestone, the extraction of which would be likely to result in loss of wildlife habitats and popular amenity; in the UK much limestone quarrying takes place in National Parks. Although there may be some market for FGD by-product gypsum, the very large amounts of material produced, if all current UK coal-fired power stations were fitted with FGD, would create a large waste disposal problem.

There are several ways for restricting vehicle emissions. In the medium to long term this could be achieved by replacing conventional hydrocarbons by less polluting fuels such as methanol, compressed or liquid natural gas, hydrogen (assuming the on-vehicle storage problem can be

* There are two main areas of environmental damage linked with acid precipitation and photochemical emissions. Acid precipitation is a cause of surface water acidification which has been observed increasingly in areas of hard rock surface cover, such as in Norway and Sweden; it is linked with the widespread loss of freshwater fisheries. Both acid precipitation and photochemical oxidants have been implicated as contributory causes to the damage to forests, particularly in Central Europe. Public concern over these two areas of environmental damage has been a potent political factor behind the drive for more extensive environmental controls relating to acidic precipitation and photochemical pollution.

solved), or electrical powered vehicles. More immediate action requires changing the design of internal combustion engines so as to minimize pollutant formation (an objective of the so-called lean burn engine), or the removal of pollutants from the vehicle exhaust gases. The latter approach requires the use of precious metal catalysts attached to the vehicle exhaust involving the installation of electronic ignition controls and the use of lead-free petrol.

The lean burn approach was favoured for a period by the UK Government who saw it as a cheaper, more robust and potentially more energy-efficient option than exhaust catalysts, and better suited to the interests of the UK car manufacturing industry (Boehmer-Christiansen, 1990, 1991). However, the failure of the lean burn engine to give satisfactory emission levels, especially of NO_x, has forced the UK to agree to the adoption of exhaust catalyst systems in order to meet EC regulations (EEC, 1989).

The initial reluctance of the UK to install FGD in existing coal-burning power stations and to adopt vehicle catalyst pollution control systems, reinforced the perception of the UK as the 'dirty man of Europe'. In both instances the UK position reflected an apparently reasonable assessment of the scientific uncertainties of the links between emissions and environmental damage (there is now acceptance that some damage has occurred (White Paper, 1990); it seemed that changing engine design to reduce pollutant formation in the combustion stage offered a potentially more robust solution to the vehicle emission problem than removing pollutants from the exhaust gases after their formation. However, it could be argued that the UK did not pay sufficient attention to the growing public and political demands in the economically advanced countries for increased environmental standards. It is likely that the UK would have benefited from devoting more effort to developing technological solutions to meet these demands rather than trying to unravel the scientifically complex cause and effect relationships so as to achieve ideal 'cost effective' solutions. If such a policy had been followed it would have improved the position of UK industry to exploit at an early stage the opportunities provided by the demand for increased environmental standards.

14.4.3 THE GREENHOUSE EFFECT AND FUTURE CLIMATE CHANGE

This is now the subject of continuing and growing global debate (Everest, 1988; Commonwealth Group of Experts, 1989; Grubb, 1990). It is the archetypal global environmental issue with strong links to population growth, to global industrial development, energy policy and social well-being. It is an issue where some nations may gain while others lose, and where adaptation may be at least a partial policy response. The recent report of Working Group 1 of the Intergovernmental Panel on Climate Change (IPCC, 1990) must be considered as providing policy-makers with the consensus view of the international scientific community on the current state of scientific understanding in this field.

Carbon dioxide is the most important of the greenhouse gases (others are methane, nitrous oxide N_2O, CFCs and tropospheric ozone), and it is responsible for approximately 60% of a greenhouse-linked climate warming. Its concentration in the atmosphere is increasing by approximately 0.5% a year, reflecting carbon dioxide emissions from fossil fuel consumption (5.4 ± 0.5 GtC per year), with an additional contribution from deforestation of 1.6 ± 1.0 GtC per year; forests are natural sinks for atmospheric carbon dioxide (IPPC, 1990). Without practical methods for removing carbon dioxide from combustion exhaust gases and then disposing of it, reduction of future emissions must depend on less use of fossil fuels. This can be achieved either by increased energy use efficiency or by changes in energy supply.

Efforts to increase efficiency are directed at reducing the energy used in the production of goods and provision of services, including more energy-efficient commercial and domestic buildings, appliances and vehicles. Improvements might be assisted through appropriate regulatory measures and economic instruments, possibly backed by socio-economic measures such as the substitution of public for private transport (White Paper, 1990, incuding annex C). The achievement of increased energy use efficiency throughout the UK economy has been an objective of the Government for many years, and from 1983 this task has been coordinated by the Energy Efficiency Office (EEO), now attached to the Department of Environment. Since its establishment the EEO has spent £130 million on programmes to stimulate better energy management, which has resulted in continuing energy savings worth £500 million a year; in 1989 the UK produced 25% more GDP than in 1979 for the same amount of energy (White Paper, 1990).

Most commentators and governments favour increased energy efficiency as a policy response to the greenhouse effect-climate change issue. It is favoured by the environmentally aware because of the emphasis it places on the decreased use of natural resources and thus on increased 'sustainability'. It is favoured by industry because it reduces the contribution made by energy consumption to manufacturing costs, and also because it reduces pressures to adopt what are perceived as less desirable measures, such as increases in energy taxes. It is favoured by governments because of its perceived benefits for employment, balance of payments, security of supply and also because it reduces the immediate need to take potentially more unpopular measures. However, an increase in energy efficiency results in a decrease in the energy cost of a given unit of output, which is equivalent economically to a reduction in the cost of energy. This means that at least part of the gains in energy efficiency will go to increase output rather than to reduce energy demand.

In the shorter term changes in energy supply wil be concentrated on the substitution of natural gas for high carbon fuels such as coal: natural gas produces only about half the carbon dioxide per unit of energy output compared with coal. However, natural gas combustion produces some carbon dioxide supply constraints may emerge, so that in the longer term an expansion in the proportion of non-fossil fuel energy sources, such as nuclear power and renewable energy technologies must take place. This would have a bonus that acid rain and photochemical pollution would be reduced, or even eliminated.

Non-fossil fuel energy sources have their own environmental problem. For example, although both nuclear and hydropower could make major contributions to reducing carbon dioxide emission, their expansion would be subject to public and political opposition. This is particularly strong for nuclear power, based on the perceived environmental, safety and social impacts of this technology and its links to military applications. Reaction against the large-scale development of hydropower is strongest at the local level because its high land use intensity leads to the loss of both human and wildlife habitats. Similar concerns are likely to be encountered when the large-scale expansion of other renewable energy sources becomes an immediate policy option, e.g. with tidal, wind or solar technologies, or the cultivation of special biomass crops for direct combustion or conversion into liquid fuels such as ethanol. Means to face and overcome these concerns must be found, if carbon dioxide emissions are to be restricted.

At present it is the richer economically advanced countries and the centrally planned economies who are responsible for the majority of global carbon dioxide emissions. However, the growing populations of the lesser developed countries and the need to improve the conditions of their often very poor citizens will lead inevitably to an increase in their energy consumption. This will almost certainly be based on current fossil fuel-based technology, as these countries will find it difficult to find the required financial, technical or land resources for the large-scale development of non-fossil fuel energy technologies. This will lead to a rise in carbon dioxide emissions from these countries which will have to be balanced by a proportionately larger decrease in emissions from the industrialized countries. It is perhaps significant that policy-makers in some of the industrialized countries are starting to make emission reduction commitments. For example, the UK aims to stabilize its carbon dioxide output by 2005 at the 1990 level (White Paper, 1990).

How any global carbon dioxide emission reduction targets will be divided between the present industrial countries and the lesser developed countries, and the policing of these targets, is likely to be a central item of debate in international negotiations on the greenhouse effect–climate change issue. To achieve agreement the carbon dioxide reduction targets for the economically advanced countries must avoid such large economic penalties as would damage their wealth-generating capacity and lose the support of their populations. Equally, any acceptable carbon dioxide emission reduction targets for the lesser developed countries must not be perceived to restrict their ability to develop economically.

A reduction in the rate of deforestation, or even its reversal, is justified both on ecological grounds and for making a contribution to stabilizing atmospheric carbon dioxide levels. However, this policy is likely to prove more appealing to the economically advanced countries, who have already destroyed much of their original forest cover, than it will be for the developing countries. The latter see the timber and land resources of their forests as essential for their economic development and one that can be exploited quickly to meet immediate problems.

Increased scientific understanding of the greenhouse effect–climate change issue will improve the assessment of adaption as a response strategy. If climate warming is at the higher end of current predictions, then some measure of adaptation will have to be part of any policy response, whilst if the warming is small adaptation might be the best overall policy. Indeed, including some adaptive measures in any response strategy could lead to policies which have lower economic penalties, and thus a greater chance of winning public support. An increase in scientific understanding also would allow balanced judgements to be made of the comparative risks associated with a future climate warming as compared with the risks associated with the large-scale expansion of alternative non-carbon dioxide emitting energy sources. It might also help to reconcile differences in the perception on the greenhouse issue between different countries and regions, and those in different stages of economic development.

14.4.4 WORLD POPULATION TRENDS

World population is in a period of unprecedented growth, having increased from approximately two and a half billion in 1945 to the present-day figure of five billion and an anticipated eight billion by 2020–30, eventually levelling out at a figure between seven and fourteen billion. Some 45% of the world's population already lives in cities, with all the environmental and social problems associated with large conurbations, and this percentage is expected to grow to 70% by 2030. Concentrations of population lead to concentrations of industry, to increased traffic density, and increasing industrialization and urbanization lead to growing demands for electricity and other fuels (Roberts *et al.*, 1990). These changes in population distribution will therefore exacerbate other environmental problems such as air and water pollution and waste disposal and, when coupled with the legitimate demands of this growing population for reasonable living standards, may make these and other environmental problems nearly impossible of solution along presently envisaged lines.

This unrestrained population growth is probably the most important issue facing the world's governments and provides their most intractable ethical problem. Indeed the solution of the environmental and social problems facing global society ultimately depend on the solution of the population growth problem. One way forward would be to concentrate effort on achieving faster global economic growth, especially in the lesser developed countries. Not only will this provide the resources to meet the demands of emerging environmental and social issues, but increasing living standards has proved to be the only effective and popularly acceptable way of limiting population growth. It may provide a more practical approach to the environmental and social problems faced by society than the low consumption paths often advocated by the environmentally concerned.

REFERENCES

Annual Review of Government Funded Research and Development (1989) Cabinet Office, HMSO, London.

Boehmer-Christiansen, S.A. (1990) Vehicle Emission Regulation in Europe – The Demise of Lean-Burn Engines, the Polluter Pays Principle . . . and the Small Car?, *Energy and Environment*, **1**, 1–25.

Boehmer-Christiansen, S.A. (1991) British Decision-Making on Vehicle Emissions: the Emerging Environmental Dimension, *Energy and Environment*, **1**, 282–306.

Brackley, P. (1987) *Acid Deposition and Vehicle Emissions: European Environmental Pressures on Britain*, Royal Institute of International Affairs, London.

COMARE (Committee on Medical Aspects of Radiation) (1986) *The Investigation of the Possible Increased Incidence of Cancer in West Cumbria*, HMSO, London.

COMARE (1988) *Investigation of the possible increased incidence of leukaemia in young people near the Dounreay Nuclear Establishment, Caithness, Scotland*, HMSO, London.

Commonwealth Group of Experts (1989) *Climate Change: Meeting the Challenge* Report of the CGE, Commonwealth Secretariat, London.

Davies, D.J.A. *et al.* (1990) Lead uptake and blood lead in two-year-old UK urban children. *Total Environ.*, **90**, 13.

Department of the Environment, *Stratospheric Ozone Review Group*, first report 1987, second report 1988, HMSO, London.

Douglas, M. (1987) *Risk Assessment According to the Social Sciences*, Russell Sage Foundation, Chicago.

Economic Community (1988) Directive 88/609/EC on *Large Plant Emissions*. This directive requires collective sulphur dioxide emission reductions of 20% by 1993, 40% by 1998 and 60% by 2003 based on 1980 emissions as baseline. Brussels, Luxembourg.

Economic Community (1989) Directive 89/458/EC on *Emissions from Motor Vehicles*, Brussels, Luxembourg.

Everest, D.A. (1988) *The Greenhouse Effect: Issues for Policy Makers*, Royal Institute of International Affairs, London.

Everest, D.A. (1990) The provision of expert advice to Government: the role of advisory committees. *Sci. Public Affairs*, **4**, 17–40.

Grubb, M.J. (1990) *Energy Policies and the Greenhouse Effect, vol. 1 Policy Appraisal*, Royal Institute of International Affairs, London.

(IPCC) Houghton, J.T., Jenkins, G.J. and Ephraums, J.J. (1990) *Climate Change: The IPCC scientific assessment*, Cambridge University Press, Cambridge.

Medical Research Council (1985, 1988) *The Neuropsychological Effects of Lead in Children – A Review of Recent Research 1979–83 and 1984–1988*, London.

Quinn, M.J. and Delves, H.T. (1989) The UK blood lead monitoring programme 1984–87: results for 1986. *Hum. Toxicol.*, **8**, 205.

Roberts, L.E.J., Kay, R.C. and Wilkinson, A.J. (1990) Urbanisation and land planning: causes and effects of climate change, in *Proceedings of Conference on Global Change: Effects on Tropical Forests, Agriculture, Urban and Industrial Ecosystems*, Bangkok, Thailand.

Royal Commission in Environmental Pollution (1983) *Lead in the Environment*, 9th Report of the RCEP, Cmnd 8852, HMSO, London.

Royal Commission on Environmental Pollution (1989) *The Release of Genetically Engineered Organisms to the Environment*, 13th Report of the RCEP, Cm 720, HMSO, London.

Rutter, M. (1983) Low level lead exposure: sources, effects and implications, in *Lead versus Health: Sources and Effects of Low Level Lead Exposure* (eds M. Rutter and R. Jones), Wiley, Chichester.

Weinberg, A. (1972) *Science and Trans-Science. Minerva*, **10**, 209–213 (especially section on axiology in science).

White Paper (1990) *This Common Inheritance: Britain's Environmental Strategy*, CM 1200, HMSO, London.

Environmental concern 15

R.J. Berry

Sam Berry has been Professor of Genetics at University College London since 1978. He was President of the Linnean Society, 1982–85, of the British Ecological Society, 1987–89, and of the European Ecological Federation, 1990–92. He wrote the ethics section of the UK response to the World Conservative Strategy, and was one of the UK delegates to the 1989 Economic Summit Nations Conference on Environmental Ethics, after which he chaired a Working Party for the European Commission (hosts of the 1989 Conference) which produced a 'Code of Environmental Practice' (see Appendix).

15.1 INTRODUCTION

Decisions become harder as options become restricted. As more and more of us are born and begin to draw on the resources of our finite world, we are faced with increasingly difficult problems. Until comparatively recent times the world could be, and was, treated as effectively inexhaustible. Environmental problems were local ones caused by crowding or over-exploitation, and solvable by migration or fairly obvious regulation. This is not to deny that some of our forebears failed to avoid environmental disasters. The early human colonists of New Zealand ate out their most convenient food, the flightless Moas; the ancient Babylonian civilization

Environmental Dilemmas Ethics and decisions
Edited by R.J. Berry
Published in 1992 by Chapman & Hall, London. ISBN 0 412 39800 1

declined through salinity levels produced by an over-extended irrigation network; the Easter Islanders could not escape from their own mistreatment of their land. But these were local problems produced by local people. We are now entering into a situation where our actions affect those far away in time and space. Pesticides may affect non-target organisms many years after their original application; radiation and toxic chemicals may spread far from their source; apparently non-toxic compounds like PCBs and CFCs cause unexpected damage to unrelated systems. No longer can we rely sensibly on symptomatic response to environmental dilemmas. This chapter is about the need to be responsible stewards as well as competent technicians or managers in our environmental attitudes. It is a difficult lesson to learn; we are always going to react more strongly to events in our own backyard than to disasters in other people's.

An additional difficulty is that environmentalists have too often damaged their credibility by strident predictions of imminent disaster. One of the first coordinated environmental initiatives of modern times was a conference of all USA Governors called in 1908 by President Theodore Roosevelt, which produced a 'Declaration of Principles', arguing that some order was needed in the management and use of natural resources. Two years later, Gifford Pinchot, Head of the US Forestry Service proclaimed 'We have timber for less than 30 years, anthracite coal for about 50 years . . . Supplies of iron ore, minerals, oil and natural gas are being rapidly depleted'. A fine beginning was marred by crying 'Wolf' too soon and too loudly.

As far as the UK is concerned, we have a history of steadily increasing care and legislation for our environment. The General Enclosure Act of 1845 can be regarded as the beginning of modern conservation legislation, formally recognizing that enclosure was the concern of all local people, and not only the lords and commoners (i.e. those who had grazing, fishing or fuel-cutting rights). It laid down that the health, comfort and convenience of local people should be taken into account before any enclosure was sanctioned. The 1845 Act was followed by the setting up in 1865 of the Commons, Open Spaces and Footpaths Preservation Society, formed to resist attempts to enclose common lands in and around London (in particular) for building purposes; it is our oldest amenity society.

In 1893 the activities of the Society led to the establishment of the National Trust as a land company to buy and accept gifts of land, buildings and common rights for the benefit of the nation. By 1912, the National Trust owned 13 sites of special interest to naturalists, including Wicken Fen in Cambridgeshire and Cothill in Berkshire. However, such potential native reserves were acquired randomly, with apparently little regard for the national significance of their animals and plants; as a response, Charles Rothschild (second son of the first Lord Rothschild) and his associates formed a Society for the Promotion of Nature Reserves (SPNR) 'to preserve for posterity as a national possession some part of our native land, its fauna, flora and geological features'. In fact, a primary aim was to persuade the National Trust to create nature reserves. The first major achievement of the SPNR was a schedule of areas of the UK considered worthy of preservation. This listed 284 potential reserves, with their special interest noted; it was submitted to the Board of Agriculture in 1915, and is remarkably similar to those in the Government White Papers of 1947 and 1949, which prepared the way for a statutory Nature Conservancy.

The formation of the Nature Conservancy in 1949 freed the SPNR (which became the Royal Society for Nature Conservation in 1981) from being a pressure group, permitting it to concentrate on coordinating and stimulating local action, and it now serves 48 County or Regional Conservation Trusts.

The 1950s and 1960s were marked by an increasing awareness that scrupulousness in environmental housekeeping was sooner or later going to be required, and a dawning discomfort about inequities between rich and poor nations. This awareness was encapsulated by Charles Birch, an Australian zoologist:

> Originally a unit of population was simply a human being whose needs were met by eating 2500 calories and 60 g of protein a day. Man's daily need of energy was equivalent to the continuous burning of a single 100 watt bulb. A unit of population in the developed world today consists of a human being wrapped in tons of steel, copper, aluminium, lead, tin, zinc and plastic, gobbling up to 60 lbs of raw steel and many other pounds of other materials. Far from getting these things in his homeland, he ranges abroad much as a hunter and more often than not in the poorer countries. His energy need . . . is equivalent to ten 1000 watt radiators continuously burning.

15.2 FROM 1970 ONWARDS

The early 1970s were the peak of rational environmental doom and represent a watershed in conservation concern in the UK. A series of conferences in 1963, 1965 and 1970 on the 'Countryside in 1970' involved the leaders of nearly all national environmental groups, representatives of farming and landowning interests, and key industrialists and government officials. A major theme was that, with the industrialization of agriculture and the increasing use of the countryside, measures to conserve wildlife populations could no longer be confined to nature reserves.

The conferences raised consciousness of environmental problems to a new level. In 1962, Rachel Carson's book *Silent Spring* had drawn attention to the insidious dangers of persistent pesticides (research in the UK was probably more advanced than in North America at that time; Monk's Wood Experimental Station with a remit to investigate the ecological effects of pesticides was opened in 1961) and in 1967 the wreck of the Liberian oil tanker *Torrey Canyon* off Land's End alerted the public to the ever-present risks of oil pollution.

Spurred by these happenings, a Royal Commission on Environmental Pollution was set up in 1970; it is still the only standing Royal Commission on Science. It has published influential Reports on (among other topics) Estuarine Pollution (1972), Nuclear Power (1976), Agriculture and Pollution (1979), Lead in the Environment (1983), Managing Waste (1985) and the Release of Genetically Engineered Organisms into the Environment (1989).

In 1972 a computer simulation carried out at the Massachusetts Institute of Technology (MIT) was published under the title *The Limits of Growth*. Its message was that the economic and industrial systems of affluent countries would collapse about the year 2100 unless two conditions were fulfilled before then: that birth rate should equal death rate, and that capital investment should equal capital depreciation. If these criteria were met, a 'stabilized world model' could result.

The MIT model was taken as a basis for a 'Blueprint for Survival' issued in the magazine *Ecologist* (1972), and endorsed by a group of leading ecologists. Its argument was that the non-renewable resources which provided the raw materials and energy generation for much of industry were threatened with drastic depletion within a time-span that ordinarily commands political attention, as a result of exponential increase in use and population growth; and the waste which accompanies this exploitation endangers the processes which sustain human life. The authors of the manifesto proposed a radical reordering of priorities, with industrial societies being challenged to opt for stability through actions producing minimum disruption of ecological processes, maximum conservation of materials and energy, and static populations. *The Times* titled its first leader on 14 January 1972, 'the prophets may be right'.

But the calculations of the *Limits to Growth* and the *Blueprint* were rendered irrelevant within a few years by the Arab–Israeli wars and a massive increase in the price of fossil fuels. Lord Ashby (who had been the first Chairman of the Royal Commission on Environmental Pollution) took a 'Second Look at Doom', pointing out the ominous instability of ecosystems made by humans in comparison with the built-in counterweights of natural ecosystems (Ashby, 1975). He argued that 'if we experience a shift in the balance of economic power between nations which own resources and nations which need those resources to keep their economies going, one sure consequence would be an increase in the tension with social systems on both sides . . . The tempting way to resolve these tensions is by autocracy and force'. In other words, the polite, rational consensus over the use of resources was at an end. Conservation was on the international agenda but it required a change of attitudes as well as an intellectual assent to impending problems.

15.3 THE WORLD CONSERVATION STRATEGY AND BEYOND

Environmental problems did not go away with the destruction of the pre-1970 premises. A human population growing by 180 people a minute, with increasing expectations has to face uncomfortable choices about its future. In 1980, the United Nations Environmental Programme, the International Union for the Conservation of Nature, and the World Wildlife Fund (now the Worldwide Fund for Nature) issued a 'World Conservation Strategy' (WCS, 1980), linking successful development with sound conservation. Implicit in it was the concept of 'sustainable development', a theme taken up and expanded in *Our Common Future*, the Report of the World Commission on Environment and Development (1987), the 'Brundtland Report'.

The explicit aim of the WCS was set out in three objectives:

1. To maintain essential ecological processes and life-support systems;
2. To preserve genetic diversity;
3. To ensure the sustainable utilization of species and ecosystems.

The achievement of this aim was assumed to be inevitable, once the problem and possible solutions were defined. This was a major fallacy; right decisions do not automatically spring from accurate knowledge. This is well illustrated by the history of clean air legislation. The association between air pollution and death rates was established by John Graunt as early as the mid-seventeenth century. During the nineteenth century there were repeated attempts to pass clean air laws in the UK Parliament, but it was not until the London smog of 1952 led to the abandonment of *La Traviata* at Sadlers Wells and the collapse of prize cattle at the Smithfield Show that comprehensive smoke control legislation was accepted (Chapter 4).

The flaw in the WCS was quickly recognized. The Strategy, being in part a UN document, required responses from member nations of UNEP. The UK response was composed of reports from seven groups, dealing with industry, city, countryside, marine and coastal issues, international policy, education, and ethics (*The Conservation and Development Programme for the UK* (1983)). The originality in this exercise was the setting up of a group on ethics. The task of this group was to put forward practical proposals about the shaping of sensible attitudes towards the environment in the multidisciplinary no-man's-land where philosophy, psychology, politics, biology and economics meet. The group dealing with education called its report 'Education for commitment', but something more was needed. I was commissioned to produce the Ethics Report, guided by a Review Group chaired by Lord Ashby and appointed by a national coordinating committee.

The Review Group met only once. It was split, apparently irrevocably, between managers and those who regarded our environmental plight as wholly the fault of human crassness. At the time it seemed pointless in pursuing this debate. I developed an aphorism that 'we are both a part of nature and apart from nature'. This formed part of our Report which was written by me with considerable help from Lord Ashby and individual discussion with other members of the group. It would be good to think that this aphorism (or rather, the truth on which it is based) helped to defuse the polarization in environmental attitudes, at least in the UK where environmental debates have been much more rational and non-confrontational than in some countries.

It is not true that ethics are not mentioned in the WCS, but the only reference is without elaboration or justification: 'A new ethic, embracing plants and animals as well as people, is required for human societies to live in harmony with the natural world on which they depend for survival and well-being'. This indifference was criticized at a conference held in Ottawa in 1986 to review progress in implementing the Strategy, and it was resolved to include ethics in revisions of the Strategy (Jacobs and Munro, 1987).

The UK Response on ethics began with the need to determine the factors that determine attitudes, which is where the need for ethics came in; not as a branch of academic philosophy, but in the fundamental sense as an expression of moral understanding 'usually in the form of guidelines or rules of conduct, involving evaluations of value or worth'.

Value was a key concept, but determining value in the environmental sense is confusing as at least four different criteria can be applied:

1. Cost in the market-place, quantified as money;
2. Usefulness for individuals or society;
3. Intrinsic worth, which depends on the objective quality of the object valued, in contrast to the market-place cost (which is quantifiable only in relation to the price of other things that can be acquired in its place);
4. Symbolic or conceptual, such as a national flag or liberty.

These four meanings can change independently for the same object. For example, water in a river in highland Scotland or lowland England will be valued differently by an economist, since its usefulness will depend on if (1) it is to be drunk, fished, or treated as an amenity, (2) it is an object of beauty or a stinking sewer, (3) it acts as a boundary between counties or countries, (4) it forms a barrier to pest spread; and so on.

Now, our interest in and therefore valuation of the environment includes self, community and future generations, but nature itself also has its own interest in survival and health. The first three of these interests are clearly anthropocentric; they are the basis of the 1980 World Conservation Strategy. Although they may conflict with each other, in principle some accommodation is usually possible. Considerable advance has been made in recent years by many economists recognizing that proper accounting involves taking into their equations both non-material and trans-generational values.

Nature's intrinsic worth is more difficult to justify from a human point of view. The commonest rationalization is explicitly utilitarian: that we should preserve as many species as possible in case they are useful to us humans (e.g. as a source of anticancer drugs, or the elusive elixir of eternal youth). Ashby has argued that we should learn to value a landscape or a biological mechanism in the same way that we are prepared to protect and pay for human artefacts like buildings or paintings (Ashby, 1978). Bryan Norton, an American philosopher, has developed a 'weakly anthropocentric' approach, based on the proposition that we are continually being transformed by our contact with the world around us, which is therefore an integral part of our human development (Norton, 1987).

The alternative argument, that nature has 'rights' indistinguishable from those of humans, has adherents from hard philosophers to green mystics. However, it stretches credibility (although this is not necessarily a bad thing), and is alien (and ironically, antagonistic) to the notion of stewardship. Stewardship is the traditional approach to the environment, in western nations at least, although it is often defined too narrowly; a common error is to equate it with preservation rather than management for particular ends.

A revised WCS was published in 1992 under the title *Caring for the Earth*. The second WCS is an expansion of the themes set out in the Brundtland Report, *Our Common Future*, whose emphasis in turn was foreshadowed by the Environment Ministers of the Economic Summit Nations, who declared in 1984:

> We stress the importance of sustainable development, prevention rather than cure; environmental impact assessment; setting environmental standards on the basis of best technology; and development of less polluting and most cost-effective technologies . . . Environmental policy should be integrated fully into other policies.

But the problem remains: how can intellectual commitment be translated into effective action? In 1984, the Economic Summit Nations established, at the instigation of Japan, a series of annual bioethics conferences, reporting directly to the Heads of State meeting. The first five of these conferences were concerned with different aspects of medical ethics; the sixth conference, hosted by the European Commission in Brussels in May 1989, was on 'Environmental ethics: man's relationship with nature, interactions with science'. This Conference discussed a range of specifically scientific questions that could improve the quality of environmental decision-making (such as access to data collected for defence purposes, improvement of environmental monitoring, ecological consequences of large-scale deforestation), and (as the final communiqué said) 'benefited from a high degree of convergence between people of different cultures, East and West, and a wide variety of disciplines'. A recurring theme was the need for an Environmental Code of Practice. Jacques Delors, President of the European Commission, called for such a code in his opening address. He argued that both Christian and Oriental religions have failed to prevent 'environmental appropriation' and that:

> the values which have been accepted up to now by all industrial societies, whereby our natural habitat has become no more than a commodity, must be replaced by different values and a different approach to the environment.

The Brussels Conference set up a Working Party from among its members to devise a code. This code was published in the following year (Norwegian Research Council for Science and the Humanities (1990)), and is set out in full in Appendix 15.A.

15.4 THE BRUSSELS CODE

To be generally acceptable, any code of practice has to be based on widely agreed propositions. The code prepared by the Brussels Working Party was based on a simple ethic: *stewardship of the living and non-living systems of the earth to maintain their sustainability for present and future, allowing development with equity*. Health and quality of life are ultimately dependent on this.

This statement depends on several axioms involving a range of moral responses, and leads to a series of obligations.

The value of a code of practice as opposed to statutory regulation or a Charter of Rights is that issues can be considered on their own merits, rather than subsumed under a set of general prohibitions. It also removes the danger that achievement (or worse, promulgation without implementation) of particular environmental 'standards' is equated with solving the problem in question. This is well illustrated by debates about permissible emissions of radionuclides or discharges into the North Sea. On the other hand, law is necessary to protect responsible stewards from greedy entrepreneurs, and may provide a stimulus to developing environmentally-friendly technology (e.g. non-persistent pesticides or chloroflourocarbon (CFC) replacements in refrigerators) by banning (or rigorously controlling) undesirable practices.

15.5 ON FROM 1992

There are signs that an environmental ethic would be welcomed by many. In her testimony of environmental conversion to the Royal Society, Mrs Thatcher claimed that 'sustainable economic development can be achieved throughout the world provided the environment is *nurtured and safeguarded*' (my italics). The International Chamber of Commerce (ICC, 1974) issues *Environmental Guidelines for Wold Industry*, stating *inter alia* that 'industry should, in addition to the usual elements, take into account the impact on the human environment, the vulnerability of natural ecological systems, and the challenge created by the finite character of the Earth's non-renewable resources'. These guidelines have been adopted by many firms and industries as a base to company (or industry) philosophy or practice, usually developed in more detailed ways. (For example, the European Chemical Industry Federation produced *Guidelines for the Protection of the Environment* in 1987, and a code of practice for industrial waste management in 1988.) This process stimulated the formation in 1986 of an Institute of Business Ethics, which has produced a model code urging the publication by all companies of a statement about their responsibility to the environment of both employees and the general public.

The ICC is one of the few major international organizations to have accepted the concept and challenge of sustainable development:

> We should seize the business opportunities offered by green consumerism, recycling, waste minimization and energy efficiency, and at the same time show corporate responsibility and commitment of a high order in reducing the strain on the environment and in developing innovative solutions.

It has urged the practice of environmental auditing, because

> . . . if properly applied, self-regulation is frequently more effective than reliance on legislation and official regulators.

ICC interest and work in this area has continued with the production of a 'Business Charter' (see pp. 222–4), which, like the Environmental Guidelines has been accepted by many major firms.

In September 1989, the US Coalition for Environmentally Responsible Economics Project produced ten Valdez Principles, as a response to the massive *Exxon Valdez* oil spill in Alaska, asking industries to subscribe to them on a voluntary basis. The Principles cover: protection of the biosphere, sustainable resource use, waste management, energy efficiency, risk reduction, the environmental impacts of products, damage compensation, disclosure of hazards, the appointment of environmental directors, and regular environmental audit. The standards laid down would be used in self-assessment and compliance monitoring by interested parties.

In May 1990 a ministerial conference in Bergen, one of the main follow-ups to the Brundtland Commission Report, received reports from scientific and other preparatory groups, some of which urged the adoption of codes along the lines of the one developed by the Brussels group. The Norwegian Prime Minister, Jan Syse, argued strongly for 'the enforcement of a binding environmental code of conduct' as a basis for multilateral environmental monitoring at the national level.

Meanwhile, in September 1990 the UK Government issued its environmental policy as a White Paper. Remarkably for a Government document, it began with a moral principle, that 'The starting policy for this Government is the ethical imperative of stewardship which must underlie all environmental policies. Mankind has always been capable of great good and great evil. This is certainly true of our role as custodians of our planet'.

This stewardship was to be worked out through a number of supporting responses:

1. Policies must be based on fact rather than fantasy, and the best evidence and analysis available must be used;
2. Given the risks to the environment, precautionary action must be taken where it is justified;
3. Public debate and public concern must be informed by publication of the facts;
4. Work for progress in the international arena is as important as that done at home;
5. Care must be taken in choosing the best instruments to achieve environmental goals.

Part of the input to the Government in its preparation of the White Paper was a review of the Recommendations in the UK Response to the World Conservation Strategy. In reviewing the ethics section, I concluded 'an acceptable basis for an environmental ethic is now emerging, based on stewardship and sustainability'. The Brussels Code is one of the sources for the Rio Declaration promulgated at the United Nation Conference on the Environment and Development (UNCED) in June 1992. Another source was a document from the world-wide Anglican Communion, summarizing documents produced by individual churches (including the Church of England and published by the Board for Social Responsibility of the General Synod as *Christians and the Environment*, G S Misc. 367, 1991) (Appendix 15.B). The encouraging trend is the convergence taking place between the religious and secular, the academic and managerial, and governmental and non-governmental agencies. Environmental dilemmas will never disappear, but at least we seem to be approaching agreement on a framework for dealing with them.

15.6 ACKNOWLEDGEMENTS

This chapter is an expansion of a 1990 paper, 'Environmental knowledge, attitudes and action: a code of practice', published in *Science and Public Affairs*, **5**, 13–23. The Brussels 'Code of Practice' was originally published in *Sustainable Development, Science and Policy* by the Norwegian Research Council for Science and the Humanities, Oslo (1990, pp. 345–57).

BIBLIOGRAPHY AND KEY REFERENCES

Ashby, E. (1975) *A Second Look at Doom*, Twenty-first Fawly Foundation Lection, Southampton, University of Southampton.

Ashby, E. (1978) *Reconciling Man with the Environment*, London, Oxford University Press.

Ashby, E. and Anderson, M. (1981) *The Politics of Clean Air*, Oxford, Clarendon Press.

Attfield, R. (1991) *The Ethics of Environmental Concern*, Oxford, Basil Blackwell.

Bourdeau, Ph., Fasella, P.M. and Teller, A. (eds) (1989) *Environmental Ethics. Man' Relationship with Science*, Brussels, CFC, EUR 12848EN.

Caring for the Earth. A Strategy for Sustainable Living (1991) Gland, Switzerland, The World Conservation Union.

Engel, J.R. and Engel, J.G. (eds) (1990) *Ethics of Development: Global Challenge, International Response*, London, Belhaven.

Environmental Guidelines for World Industry (1974, revised 1986, 1990). Paris, International Chamber of Commerce.

Group of Legal Experts Advisory to the World Commission for Environment (1988). *Legal Principles for Environmental Protection and Sustainable Development*, Dordrecht, Martin Nijhoff.

Jacobs, P. and Munro, D.A. (eds) (1987). *Conservation with Equity. Strategies for Sustainable Development*, Gland and Cambridge, International Union for the Conservation of Nature.

Norton, B.G. (1987). *Why Preserve Natural Variety?* Princeton, N J, Princeton University Press.

The Conservation and Development Programme for the UK (1983) A response to the World Conservation Strategy, London, Kogan Page.

This Common Inheritance (1990). Britain's Environmental Strategy, London, HMSO, Cm 1200.

World Charter for Nature (1982) Resolution 37/7 of United Nations General Assembly, 48th Plenary Meeting, 28 October, pp. 17–18.

World Commission on Environment and Development (1987). *Our Common Future*, Oxford and New York, Oxford University Press.

World Conservation Strategy (1980) Gland, Switzerland, International Union for the Conservation of Nature.

Appendix A: A Code of Environmental Practice

BACKGROUND

We live in a world which for millions of years has supported an awesome variety of plants and animals, and a human population which is growing by 150 people a minute; it now numbers over five billion and will be six billion by the end of the century. But Planet Earth is finite, and resources will inevitably become limiting unless they are allowed to recover from overuse or misuse, or be substituted by acceptable equivalents. We are neither Earth's master nor its slave but the evidence is now overwhelming that our present behaviour towards it cannot continue. Global warming, destruction of the ozone shield, acidification of land and water, desertification and soil loss, deforestation and forest decline, diminishing productivity of land and waters, and extinctions of species and populations* show that human demand is exceeding environmental support capacities.

Environmental stress is often seen as the cost of growing demand on scarce resources and pollution generated by rising living standards in the industrialized countries, where urban growth and industrial development have replaced the pressure of rural poverty and population growth. But environmental degradation may come from poverty as well as

* This code was called for by the Sixth Economic Summit Nations Conference on Bio-Ethics, meeting in Brussels 10–12 May 1989. It was prepared by a Working Party consisting of R.J. Berry (UK), D. Birnbacher (Germany), P. Bourdeau (CEC), Abbyann Lynch (Canada) and A. Morishima (Japan).

Environmental Dilemmas Ethics and decisions
Edited by R.J. Berry
Published in 1992 by Chapman & Hall, London. ISBN 0 412 39800 1

affluence through increasing numbers of mostly landless people having to destroy their own resource base, often using up fuel wood faster than it is grown, farming marginal land at non-sustainable levels, depleting water supplies and overgrazing rangelands.

The relationships between population, resources and the environment are complex, and are complicated by inequity and inefficiency in industrialized and developing countries alike: poverty, injustice, and environmental degradation interact to create tension by competition for non-renewable resources, land or energy. Serious environmental problems are the result of both short-term expediency and longer-term ignorance, actions taken without full account or awareness of their consequences. Development cannot take place on a deteriorating environmental resource base; the environment cannot be protected when growth leaves out of account the costs of environmental destruction. Development and the aim of an acceptable quality of life for all cannot be separated from environmental management; both must be integrated with all the facilities of national and international bodies. 'This is where the ethical approach comes in; it concerns the values which govern social behaviour. It is the bedrock of law and therefore determines the various codes by which we act, codes hallowed by tradition and whose real cruxes must now be re-established. The continuous degradation of the setting for life which we have received as a legacy from our forebears will of necessity prompt us to adopt an approach to that legacy in terms of duties and responsibilities' (Jacques Delors, addressing the Environmental Ethics Conference in Brussels, 10 May 1989).

ENVIRONMENTAL PRACTICE

Former Norwegian Prime Minister, Mrs Gro H. Brundtland, has called for 'a new holistic ethic in which economic growth and environmental protection go hand in hand' (World Conference on the Changing Atmosphere, 1988). The role of an ethic is to provide inner incentives to act in ways that are ultimately conducive to the well-being of the individual or group concerned where reliable external incentives appealing to the agent's immediate self-interest are absent. To be generally acceptable, it must be independent of race, culture and religion. The value of an environmental ethic was defined by the pioneer American ecologist Aldo Leopold as 'a mode of guidance for meeting ecological situations so new or intricate, or involving such deferred reactions, that the path of social expediency is not discernible to the average individual'. Since we are dependent for survival, health and psychological well-being on the physical integrity of the biosphere as well as the aesthetic quality and cultural continuity of our local natural environment, an environmental ethic that encourages responsibility to the human use of natural resources is an instrument rather than an opponent of our self-interest as a community stretching through generations. From the perspective of our destiny as a species, such an ethic is essentially an adaptive device, functionally related to our continued thriving as part of the natural world; it is conditioned by, but not dependent upon, any particular religion and culture.

Thus, environmental ethics are more than self-interest. Obigations to protect the environment are incumbent on everyone, but responsibility, which is an increasing function of power, rests in particular with communities, governments, and corporations. Furthermore, basic justice demands that any sacrifices ought to be distributed according to capacity. This means that the main burdens of responsible action will fall on the most highly developed countries whatever their causal role may have been in making the problems exist in the first place; this does not absolve the less developed or poorer countries from accepting a proper share of the necessary costs, since all humanity is implicated.

Basic principles for an Environmental Ethic are stewardship, sustainability and quality of life. These require us to recognize that:

1. We depend on the natural world for food, fuel, space and stimulation; and the quality of life for people today and for our children tomorrow depends on how successful we are in cherishing that resource;
2. The diversity and beauty of our natural environment are immense gifts with immense inspirational qualities, and their degradation affects us both materially and spiritually.

QUALITY OF LIFE

We are dependent upon our environment, yet capable of manipulating and managing it to a variable extent; we are simultaneously apart from, yet a part of our physical, anthropogenic and social environment, which may stimulate or adversely stress us. A high quality of life is a legitimate aspiration for all, but we must distinguish quality of life from standard of living. Health is best defined positively as wholeness of all aspects of life, and not negatively as an absence of disease; health is the strength to be human, with an environment to rejoice in and respond to rather than avoid. The minimum requirement for health is the satisfaction of basic biological and social needs, but a good quality involves also the ability to accept and adjust to change.

Individuals have different priorities, desires and perceived wants, which are determined by social pressures, culture and religion. It is not possible to define more than the basic minimum of health from the environmental point of view; it is not dependent upon any specific environmental input, nor by a general lack of stress.

An effective ethic must be based on:

1. *Environmental monitoring and pathway knowledge*, by appropriate specialists and researchers so that spatially and temporally distant impacts as well as immediate ones are recognized and evaluated. The knowledge which must precede informed decision-making is likely to involve breaking traditional boundaries between scientific disciplines, since these can lead to a spurious reductionism (as distinct from the necessary reductionism of scientific experiment);
2. *Sustainable development*, defined not as a fixed state of harmony, but as a process of change in which the exploitation of resources, the direction of investments, the orientation of technological development, and institutional change are constantly readjusted to reconcile present and future needs. Sustainable development is a process of social and economic betterment that satisfies the needs and values of all interest groups, while maintaining future options and conserving natural resources and diversity. Neither market-based nor planned economies have built-in features guaranteeing substainability; neither utilitarianism nor libertarianism can by themselves support intra- and intergenerational equity for all;
3. *Full accounting of costs* to ensure that the stock of renewable resources is maintained constant, with the amount of waste or pollution kept below the assimilative capacity of the environment; and non-renewable resources either substituted or harvested minimally;
4. *A recognition of interdependent values* by individuals, communities, to future generations, and instrinsically of nature itself (at least as a factor transforming human attitudes), differing and often conflicting in their recognition. Such values must include user (instrumental) values, existence values as part of the total economic value of species or ecosystems, and calculation of the combined technical and social costs of environmental damage (including pollution abatement). An adequate weighting of all valuations is necessary to resolve conflicts, and necessarily makes use of cost-benefit analysis, willingness to pay, and other techniques. For this purpose it is not necessary to distinguish whether a high valuation for nature is based on anthropocentric usefulness, intuitive wonder at nature's power and intricacy, respect for all living things, or a combination of all three. However, both undiluted anthropocentrism and ecocentrism are inadequate;
5. *Individual and corporate stewardship*, which means an acceptance that we are trustees, curators, guardians and wardens of our environment for both present and future generations, providing accountability, responsibility and continuity. We are environmental citizens with privileges and responsibilities; we have responsibilities to those with whom we share responsibilities.

AN ENVIRONMENTAL ETHIC INVOLVES:

Stewardship of the living and non-living systems of the earth in order to maintain their sustainability for present and future, allowing development with forebearance and equity. Health and quality of life for humankind are ultimately dependent on this.

Such an ethic entails characteristics common to good citizens, states and corporations everywhere:

1. *Responsibility* – in strengthening good management and restraining present actions so that no irreversible damage is done or unacceptable risks imposed on current or future generations;
2. *Freedom* – leaving open for both society and future generations the widest possible range of environmental choices, thus making us accountable to our community and our children, and constrained to that extent in our own decisions;
3. *Justice* – for individuals of all nations, recognizing that some carry environmental burdens which may require help from the more privileged – justice also demands that we impose no deliberate burdens on future generations by our own actions;
4. *Truthfulness* – in seeking the cause of environmental problems and dealing with them in open and honest debate.
5. *Sensitivity* – to the interdependence of ecological systems on which we depend for our own survival, yet manage for our own ends; we are both a part of and at the same time apart from our environment, and our management must be based on the best and most up-to-date information possible;
6. *Awareness* – that quality of life arises from multiple interactions with our physical, social and biological environment, whether natural or anthropogenic; and that quality of life is distinct from, and only partially dependent on, standard of living;
7. *Integrity and decisiveness* – that environmental scientists and decision makers have especial responsibilities for collecting and disseminating information, for education, and for prudent environmental management.

   – that restraint may be needed where knowledge is incomplete, and that action may be needed on some issues before a scientific consensus emerges.

GUIDELINES

Acceptance of an ethic based on sustainability and stewardship means we must:

1. Maintain the ecological processes and life-support systems (such as

soil regeneration and protection, the recycling of nutrients, and cleansing of water) essential for a functional and productive biosphere;

2. Preserve genetic, species and ecosystem diversity on which the efficient, long-term functioning of natural life-support systems depend;
3. Ensure the stable regulation of species and ecosystems through integrating natural resource conservation with socioeconomic development;
4. Work towards a better understanding of the environment through research, education, and dissemination of information. This involves commitments to the support of environmental research, the maintenance and sharing of data banks, and a willingness to seek expert advice;
5. Pay particular respect to the global commons (the oceans and outer space; Antarctica can be regarded as an effective and exemplary Common while it is managed through the Antarctic Treaty system) where sustainable development can only be secured through international cooperation for surveillance, development and management in the common interest;
6. Respect an underpinning of law constraining (and sometimes defining) environmental standards and behaviour;
7. Bias decisions.
   (a) Against irreversible choices;
   (b) In favour of protection to those especially vulnerable to our actions and choices;
   (c) In favour of sustainable rather than one-off benefits;
   (d) Against causing harm, as opposed to merely forgoing benefits.

These general guidelines produce specific

OBLIGATIONS

Some of these obligations affect states or corporate bodies, others involve individual action.

1. All environmental impacts should be full assessed in advance for their probable effect on the community, posterity, and nature itself, as well as on individual interest;
2. Regular monitoring of the state of the environment should be undertaken and the data made available without restriction;
3. The provisions of support for basic environmental research as well as conservation, resource and pollution studies, to ensure and improve knowledge of environmental processes;
4. The assessment of activities involving environmental impact should incorporate social, cultural and environmental costs, as well as commercial considerations;

5. The facilitation of technological transfer, with justice to those who develop new technologies and equitable compassion towards those who need them;
6. Regulatory and mandatory restrictions should be effected wherever possible by cooperation rather than confrontation; minimum environmental standards must be effectively monitored and enforced;
7. Regular review of environmental standards and practices should be undertaken by expert independent bodies;
8. Costs of environmental damage (fully assessed as in 4. above) should be fully borne by their instigator, including newly-discovered damages for an agreed period retrospectively;
9. Existing and future international conventions dealing with transfrontier pollution or the management of shared natural resources should include:
   (a) The responsibility of every state not to harm the health and environment of other nations;
   (b) Liability and compensation for any damage caused by third parties;
   (c) Equal right of access to remedial measures by all parties concerned;
10. Industrial and domestic waste should be reduced as much as possible, if appropriate by taxation and penalties on refuse dumping. Waste transport should be minimized by adequate provision of recycling and treatment plants;
11. Appropriate sanctions should be imposed on the selling or export of technology or equipment that fails to meet the best practicable environmental option for any situation;
12. International agreement should be sought on the management of extra-national resources (atmosphere, deep-sea, and continued for the regions covered by the Antarctic Treaty system).

These obligations will be expanded by individual states, international organizations, transnational corporations, industrial and religious groupings for their own problems and concerns, with particular attention to the resolution of dilemmas and the establishment of sanctions against code violators. Governments should develop their own foreign policy needs to reflect the fact that their policies have a growing impact on the environmental resource base of other nations and commons, just as the global policies of other nations have an impact on its own.

CONFLICTS AND SANCTIONS

Adherence to an environmental ethic or code precedes and enables the enforcement of any consequent regulations. The role of such regulations is to protect the values implicit in the ethic, rather than maintain arbitrary standards. Values give rise to obligations, and thence to particular standards for behaviour. Failure to conform to these standards may be inadvertent or wilful, but any failure must be subject to both deterrent sanctions and proper reparation (i.e. to a strict enforcement of the polluter pays principle). Several states have Environmental Protection Acts, which recognize the concept of public trust and place a duty on a statutory agency, as well as businesses and developers who own natural resources, to protect them from pollution and degradation. A central element is a shifting of some of the burden of proof from the plaintiff to the accused; a plaintiff need not establish personal injury, and an alleged polluter has to establish proof of the inoffensiveness of his or her activity.

Environmental protection requires an additional role for law, beyond simple protection or sanction:

1. To create a framework for scientists and politicians to interact;
2. To facilitate public access to information;
3. To stimulate technological development by restriction on unacceptable practice.

Many disputes can be avoided or more readily resolved if these principles and responsibilities are built into legal frameworks, which are respected and implemented. Individuals and states are more reluctant to act in a way that may lead to a dispute when there are established and ultimately binding procedures for settling disputes. However, these are largely lacking at the international level. A set of legal principles proposed by a panel of experts convened by the World Commission on Environment and Development are attached to this Code; they are not part of a Code, but are complementary to it.

LEGAL PRINCIPLES

Summary of Proposed Legal Principles for Environmental Protection and Sustainable Development, produced by a Group of Legal Experts, and published by the World Commission on Environment and Development (1988).

I. GENERAL PRINCIPLES, RIGHTS AND RESPONSIBILITIES

1) *Fundamental Human Rights* All human beings have the fundamental right to an environment adequate for their health and well-being.
2) *Inter-Generational Equity* States shall conserve and use the environment and natural resources for the benefit of present and future generations.
3) *Conservation and Sustainable Use* States shall maintain ecosystems and ecological processes essential for the functioning of the biosphere, shall

preserve biological diversity, and shall observe the principle of optimum sustainable yield in the use of living natural resources and ecosystems.
4) *Environmental Standards and Monitoring* States shall establish adequate environmental protection standards and monitor changes in and publish relevant data on environmental quality and resource use.
5) *Prior Environmental Assessments* States shall make or require prior environmental assessments of proposed activities which may significantly affect the environment or use of a natural resource.
6) *Prior Notification, Access, and Due Process* States shall inform in a timely manner all persons likely to be significantly affected by a planned activity and to grant them equal access and due process in administrative and judicial proceedings.
7) *Sustainable Development and Assistance* States shall ensure that conservation is treated as an integral part of the planning and implementation of development activities and provide assistance to other States, especially to developing countries, in support of environmental protection and sustainable development.
8) *General Obligation to Cooperate* States shall cooperate in good faith with other States in implementing the preceding rights and obligations.

II. PRINCIPLES, RIGHTS, AND OBLIGATIONS CONCERNING TRANSBOUNDARY NATURAL RESOURCES AND ENVIRONMENTAL INTEFERENCES

9) *Reasonable and Equitable Use* States shall use transboundary natural resources in a reasonable and equitable manner.
10) *Prevention and Abatement* States shall prevent or abate any transboundary environmental interference which could cause or causes significant harm (but subject to certain exceptions provided for in Art. 11 and Art. 12 below).
11) *Strict Liability* States shall take all reasonable precautionary measures to limit the risk when carrying out or permitting certain dangerous but beneficial activities and shall ensure that compensation is provided should substantial transboundary harm occur, even when the activities were not known to be harmful at the time they were undertaken.
12) *Prior Agreements When Prevention Costs Greatly Exceed Harm* States shall enter into negotiations with the affected State on the equitable conditions under which the activity could be carried out when planning to carry out or permit activities causing transboundary harm which is substantial but far less than the cost of prevention. (If no agreement can be reached, see Art. 22.)
13) *Non-Discrimination* States shall apply as a minimum at least the same standards for environmental conduct and impacts regarding transboundary natural resources and environmental interferences as are applied

domestically (i.e. do not do to others what you would not do to our own citizens).

14) *General Obligation to Cooperate on Transboundary Environmental Problems* States shall cooperate in good faith with other States to achieve optimal use of transboundary natural resources and effective prevention or abatement of transboundary environmental interferences.

15) *Exchange of Information* States of origin shall provide timely and relevant information to the other concerned States regarding transboundary natural resources or environmental interferences.

16) *Prior Assessment and Notification* States shall provide prior and timely notification and relevant information to the other concerned States and shall make or require an environmental assessment of planned activities which may have significant transboundary effects.

17) *Prior Consultations* States of origin shall consult at an early stage and in good faith with other concerned States regarding existing or potential transboundary interferences with their use of a natural resource or the environment.

18) *Cooperative Arrangements for Environmental Assessment and Protection* States shall cooperate with the concerned States in monitoring, scientific research and standard setting regarding transboundary natural resources and environmental interferences.

19) *Emergency Situations* States shall develop contingency plans regarding emergency situations likely to cause transboundary environmental interferences and shall promptly warn, provide relevant information to and co-operate with concerned States when emergencies occur.

20) *Equal Access and Treatment* States shall grant equal access, due progress and equal treatment in administrative and judicial proceedings to all persons who are or may be affected by transboundary interferences with their use of a natural resource or the environment.

III. STATE RESPONSIBILITY

21) States shall cease activities which breach an international obligation regarding the environment and provide compensation for the harm caused.

IV. PEACEFUL SETTLEMENT OF DISPUTES

22) States shall settle environmental disputes by peaceful means. If mutual agreement on a solution or on other dispute settlement arrangements is not reached within 18 months, the dispute shall be submitted to conciliation and, if unresolved, thereafter to arbitration or judicial settlement at the request of any of the concerned States.

Appendix B: Christians, The Environment and Development

STATEMENTS FROM THE ANGLICAN COMMUNION

Five regions of the Anglican Communion world-wide have made policy statements about proper Christian attitudes to the environment and development (the Anglican Church of Australia, Church of England, Church of Melanesia, Episcopal Church of the USA, ECUSA, and the Philippine Episcopal Church. In addition, the Church of the Province of New Zealand has supported significant work on these issues).

The common belief of all these Churches is (in the words of the ECUSA statement):

> All creation is of God and as part of creation we are given the specific tasks of responsible and faithful stewardship of all that is.

This involves:

1. The clear understanding from both scripture and enduring tradition that:
   (a) Responsible stewardship means that we are representative caretakers, managers or trustees, accountable for our actions;
   (b) The Christian gospel of reconciliation extends from 'a change from a level of human existence that is less than that envisaged by our Creator, to one in which humanity is fully human and free to move to a state of wholeness in harmony with God, with fellow human

Environmental Dilemmas Ethics and decisions
Edited by R.J. Berry
Published in 1992 by Chapman & Hall, London. ISBN 0 412 39800 1

beings and with every aspect of his environment' (Statement of Sixth Meeting of the Anglican Consultative Council, 1987);

2. The misuse or misappropriation of the finite resources of the earth by ever increasing numbers of people is unsustainable, unjust and morally reprehensible;
3. Human dignity cannot be achieved or maintained in a degrading environment with a declining resource base. The traditional Christian aims of peace and justice must be supplemented by informed environmental care;
4. Environmental stewardship depends on the attitudes of individuals, corporations and governments. Because it involves maintaining the earth's sustainability for present and future allowing forbearance and fairness for all, current attitudes may have to change radically, particularly to manage actions with effects distant in time or place from their origin;
5. There is an encouraging convergence between the churches and secular bodies that stewardship is the basic ingredient for human survival and development, in the finite world in which we dwell. The Christian church can fully endorse the opening of the National Report of the United Kingdom to UNCED, which states:

> Mankind long believed that whatever we did, the Earth would remain much the same. We now know that is untrue. The ways we produce energy, and the rate at which we multiply, use natural resources and produce waste threaten to make fundamental changes in the world environment . . . We have a moral duty to look after our planet and hand it on in good order to future generations . . . We must not sacrifice our future well-being for short-term gains, nor pile up environmental debts which will burden our children.

Index